EVALUATION AND PREVENTION OF WATER DAMAGE TO ASPHALT PAVEMENT MATERIALS

A symposium sponsored by
ASTM Committee D-4 on
Road and Paving Materials
Williamsburg, VA, 12 Dec. 1984

ASTM SPECIAL TECHNICAL PUBLICATION 899

Byron E. Ruth, University of Florida,
editor

ASTM Publication Code Number (PCN)
04-899000-08

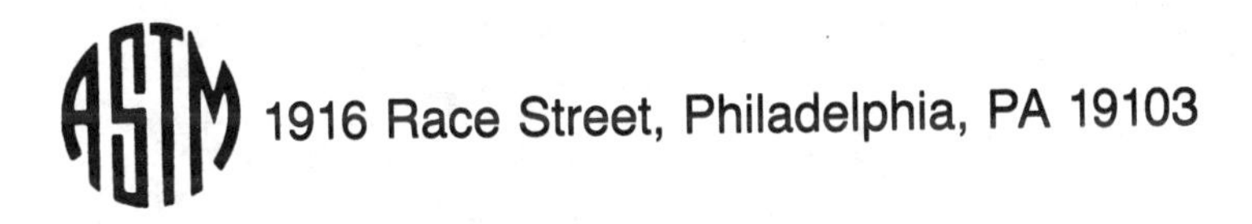

1916 Race Street, Philadelphia, PA 19103

Library of Congress Cataloging-in-Publication Data

Evaluation and prevention of water damage to asphalt pavement materials.

(ASTM special technical publication; 899)
"ASTM publication code number (PCN) 04-899000-08."
Includes bibliographies and index.
1. Pavements, Asphalt — Maintenance and repair — Congresses. 2. Road drainage — Congresses. I. Ruth, Byron E. II. ASTM Committee D-4 on Road and Paving Materials. III. Series.
TE270.E84 1985 625.8'5 85-26783
ISBN 0-8031-0460-X

Library of Congress Catalog Card Number: 85-26783

NOTE

The Society is not responsible, as a body,
for the statements and opinions
advanced in this publication.

Printed in Baltimore, MD
Dec. 1985

Foreword

The symposium on Water Damage of Asphalt Pavements: Its Effect and Prevention was presented at Williamsburg, VA, on 12 Dec. 1984. The symposium was sponsored by ASTM Committee D-4 on Road and Paving Materials. Byron E. Ruth was chairman of the symposium and is editor of this publication.

Related ASTM Publications

Pavement Maintenance and Rehabilitation, STP 881 (1985), 04-881000-08

Placement and Compaction of Asphalt Mixtures, STP 829 (1984), 04-829000-08

Properties of Flexible Pavement Materials, STP 807 (1983), 04-8070000-08

Pavement Surface Characteristics and Materials, STP 763 (1982), 04-763000-47

Asphalt Pavement Constuction: New Materials and Techniques, STP 724 (1981), 04-724000-08

Quality Assurance in Pavement Construction, STP 709 (1980), 04-709000-08

A Note of Appreciation to Reviewers

The quality of the papers that appear in this publication reflects not only the obvious efforts of the authors but also the unheralded, though essential, work of the reviewers. On behalf of ASTM we acknowledge with appreciation their dedication to high professional standards and their sacrifice of time and effort.

ASTM Committee on Publications

ASTM Editorial Staff

Contents

Contents

Overview

Historically, engineers have recognized the importance of providing good surface and subsurface drainage from pavements. Structural design either minimized the potential for water entry or allowed for a reduction in structural support resulting from the periodic variation in water table and soil moisture. In some situations it was found to be advantageous to use full depth-thick lift asphalt pavements over clay subgrades where improvement in drainage conditions was not feasible. The performance of these asphalt mixtures in the presence of moisture appears to be excellent. In fact, asphalt paving mixtures used in the facing of dams and for pond linings seem to exhibit no major problem with deterioration caused by water. However, in recent years the problems of water damage to asphalt pavement has directed attention toward the phenomenon called "stripping." This term is applied to paving mixtures that exhibit separation of asphalt films from aggregate surfaces due primarily to the action of water.

The severity and extent of stripping is primarily related to environmental conditions, materials, and adequacy of mix design and construction. In South Carolina severe stripping was observed on 8.1% of pavements sampled throughout the highway system by Bushing et al. Florida has experienced some stripping problems in the past and consequently now specifies the use of approved antistripping agents in all friction course and recycled mixtures. Stripping is most prevalent in eastern Texas where high water table or high rainfall conditions or both exist. This combined with siliceous river, rhyolite, and certain limestone aggregates that are susceptible to moisture damage tends to promote stripping unless treated with lime or other suitable antistripping agents. Oregon has found it necessary to use antistripping agents in over 20% of its construction projects since 1983.

The papers contained in this special technical publication (STP) provide considerable insight into the severity of stripping, test methods for evaluation, and relative effectiveness of using antistripping additives to minimize in-service stripping potential. However, the mechanism of stripping in precise physical-chemical terms and the effect of asphalt characteristics are not discussed. In general the authors have emphasized the effectiveness or ineffectiveness of antistripping agents as evaluated by various laboratory test methods.

The evaluation of stripping potential and effectiveness of antistripping agents are generally performed using a visual test (boiling test) or a retained strength test (for example, indirect tensile, compression, Marshall stability, and so forth). The boiling test is strictly an empirical test of limited value since the results may be influenced by asphalt viscosity, degree of boiling, and time of exposure. Furthermore, it is not rational to boil the coated aggregate since exposure conditions in pavements are not even remotely similar.

The use of retained indirect tensile strength using either the Lottman procedure or modifications of his procedure appears to be the most widely accepted method for evaluation of potential water or freezing damage or both to asphalt paving mixtures. The Lottman saturation, freezing, and thawing conditioning procedure has been considered too severe for some researchers and state highway agencies. The Tunnicliff method requires saturation of 55 to 80% without any freezing cycles and is conditioned the same as Lottman's procedure in a 60°C (140°F) water bath for 24 h. However, Lottman provides 2 h at room temperature for partial drying before testing whereas Tunnicliff maintains moisture in the test specimen using a 25°C (77°F) water bath and performs the indirect tension test using wet specimens and a 50.8 mm/min (2-in./min) rate of loading.

Gilmore et al considered the Lottman procedure insufficiently severe and increased the thawing time and number of freeze cycles to achieve greater reductions in tensile strength ratio. Increased soaking at 60°C (140°F) had little effect beyond 48 h but an increase in repeated freeze cycles with warm water soak produced continual reduction in the tensile strength ratio.

Test specimens for determination of the tensile strength ratio are usually prepared at a compactive effort to produce air-void contents in the range of 6 to 8% as an attempt to simulate field conditions. However, most investigators reporting on pavement studies indicate numerous cases where in-service air-void contents are substantially greater than 8%. Water damage and severe stripping often occur when more open graded, high air void surface mixtures are placed over dense, well compacted, asphalt concrete. Takallou et al and Kennedy point out the importance of providing both good mix design and field compaction to achieve relatively low air-void contents and a well sealed pavement. Laboratory tests indicate that when higher asphalt contents or other improvements reduce the air-void content, the susceptibility to moisture damage is greatly reduced and the effect of antistripping agents is not as pronounced.

There are a variety of other test methods that have been used to evaluate water and freezing damage. Kennedy used the Texas freeze-thaw pedestal test (an adaptation of the test developed at the Western Research Institute) to evaluate the effectiveness of different antistripping additives on fine river gravel and fine sand. This is a relatively simple test to perform, but the viscoelastic properties of the asphalt and their influence on test results have been neglected. If the same asphalt is used in all comparisons, the test results will at least be comparative. However, different asphalts may have widely varying shear moduli and viscosity at the recommended test temperatures. Therefore, it is entirely possible to achieve poor test results even though the aggregate is not susceptible to stripping.

Dynamic diametral fatigue tests were conducted by Kim et al to evaluate mix moisture content effects on compacted and Lattman conditioned specimens. Parameters derived from the diametral tests include resilient modulus, repetitions to failure, and permanent vertical compressive strain. Tests of this type are time consuming and perhaps suitable for research studies, but not desirable as a conventional test method to evaluate stripping potential and antistripping additives. The lack of density control (air-void content), particularly on the North Oakland-Sutherlin evaluation, makes interpretation of the test parameters difficult with respect to the deteriorative effects of moisture and conditioning.

The evaluation of cold-recycled mixtures by Tia and Wood using water immersion conditioning of gyratory compacted mixtures indicated that Hveem *R*-value tests were not sensitive to the effects of water on mixtures containing emulsion or foamed asphalt. However, significant reductions in Hveem *S*-value, Hveem cohesiometer value, Marshall stability, and diametral resilient modulus were observed between as compacted and conditioned test specimens. It was observed that increased curing time and compaction effort improved the water resistance of these mixtures. This investigation did not address mix durability resulting from freeze-thaw cycles. However, Castedo et al evaluated moisture and freeze-thaw damage to assess the effectiveness of admixtures in various foamed asphalt mixtures. The modified Marshall stability tests appeared to yield results sensitive to the effects of conditioning and antistripping additives. Pulse velocity measurements attained at a different number of freeze-thaw conditions seem to be related to the modified Marshall stability. These results provide some general guidelines for laboratory evaluation, but no information is given on the degree or severity of water and freezing damage to pavement constructed with cold mix.

The evaluation of lime and different liquid antistripping agents by Kennedy, Butlon, Castedo et al, Gilmore et al, and Kim et al generally indicate anywhere from marginal to substantial improvement in the mixture's resistance to water and freezing damage. Lime appears consistently beneficial provided the method of treatment at the plant is performed properly. The use of lime does require that it be considered in the mix design procedure since it acts as a mineral filler in addition to the effect of providing antistrip properties.

The advantage of using liquid antistripping agents over lime is the simplicity of adding them directly to the asphalt. Obviously, cost may be a factor, but in many instances this is relatively cheap insurance to minimize the potential for water damage.

The approach of requiring an antistripping agent to achieve a retained tensile strength not less than 70 to 75% (or minimum wet-dry Marshall stability of 85% and so forth) may be realistic depending upon the reliability of the test method, but it tends to negate the potential for improving a mix that marginally meets the minimum requirement. The problem of sometimes specifying or not specifying the use of antistripping agents in supplier furnished asphalts can lead to confusion and irate suppliers. One of seven aggregate sources in the Miami, FL, area was found to have stripping tendencies. In the evaluation of these materials the approved antistripping agents corrected this problem as well as increased the retained tensile strengths from about 80 to over 100% for friction course mixtures using these aggregates. Since there was a substantial improvement in test properties and a relatively low cost for antistripping agents (approximately $0.01/sq. yd.), it was specified for all friction course mixtures.

Durability of asphalt paving materials can be influenced by moisture in the mix. Aggregates and RAP with high moisture content may result in mixtures having excess moisture at the time of placement. If severe, this can produce a flushing (or bleeding) condition resulting in a low skid resistance surface. Laboratory tests demonstrate that residual moisture in the compacted specimen increases stripping potential. Recommendations generally require moisture contents less than 1.0% or the use of a suitable antistripping agent in cases where reduction in mix moisture content is not feasible.

The papers presented in this STP have been organized according to three major topics. It should be recognized that considerable overlap exists in the information provided by the authors. Although test technique and objectives are varied, there are numerous substantiating statements regarding the causes of water damage, the methods for evaluation, and recommendations for corrective action. Interpretation of the information for application to site specific problems or development of specifications is relegated to the reader of this publication.

Byron E. Ruth
University of Florida, College of Engineering, Department of Civil Engineering, Gainesville, FL 32611; symposium chairman and editor.

Methods for Identification and Evaluation of Stripping in Highway Pavement Systems

Herbert W. Busching, [1] *Gregg C. Corley,* [1] *James L. Burati, Jr.,* [1] *Serji N. Amirkhanian,* [1] *and Jerry M. Alewine* [2]

A Statewide Program to Identify and Prevent Stripping Damage

REFERENCE: Busching, H. W., Corley, G. C., Burati, J. L., Jr., Amirkhanian, S. N., and Alewine, J. M., **"A Statewide Program to Identify and Prevent Stripping Damage,"** *Evaluation and Prevention of Water Damage to Asphalt Pavement Materials, ASTM STP 899,* B. E. Ruth, Ed., American Society for Testing and Materials, Philadelphia, 1985, pp. 7–21.

ABSTRACT: A statewide pavement coring and testing program was conducted in South Carolina to determine the extent and severity of asphalt pavement stripping. Information from extensive and intensive coring and testing programs identified causes of stripping so that effective measures could be developed to eliminate or minimize future damage. Stripping frequency was related to the following factors: highway, aggregate source, mix type, cross-section type, traffic group, pavement age, and presence of open-graded friction courses. Stripping was pervasive but of varying severity in all soil provinces. Results of findings from indirect tensile strength tests of pavement core layers and a visual examination are given. Strength retention of moisture-conditioned pavement core layers was measured for specimen cores obtained from the outer wheel path.

KEY WORDS: pavements, flexible pavements, stripping (distillation), asphalt pavement stripping, causes of stripping, indirect tensile strength, retained tensile strength, stripping extent, stripping severity

Loss of asphalt adhesion and related moisture-induced damage have potentially costly consequences when pavement durability and serviceability are reduced. Until sustained adhesion to aggregate of different composition, temperature, and dryness becomes available, perhaps through development of more "forgiving" asphalts [*1*], it is likely that regional research on stripping will retain high priority.

A 1981 survey of each of the eleven Southeastern Association of State Highways and Transportation Officials (SASHTO) states identified, by all the states,

[1]Professor of civil engineering, graduate assistant, associate professor, and graduate assistant, respectively, Clemson University, Department of Civil Engineering, 214 Lowry Hall, Clemson, SC 29631.

[2]Assistant materials engineer, South Carolina Department of Highways and Public Transportation, P. O. Box 191, Columbia, SC 29202.

the occurrences of stripping and subsequent damage to pavements [1]. Problems have been reported in new construction, in asphaltic concrete overlays over existing flexible sections, and in asphaltic overlays over portland cement concrete pavements. Stripping is generally related to moisture, and it may therefore be hypothesized that the severity of stripping may be seasonal, with moist pavements more distressed by stripping than dry pavements. South Carolina and other southeastern states have relatively high annual average rainfall, averaging between 110 and 140 cm (43 and 55 in.) annually for South Carolina. Rainfall is distributed throughout most of the year with portions of the fall (September and October) being dryer than other months. Moisture is therefore present throughout much of the year to promote stripping.

Some of the stripping problems identified in the SASHTO states appeared to be related to the presence of open-graded friction courses [2]. Some southeastern states, including South Carolina, have imposed a moratorium on the use of plant-mixed seal courses (PMSC) and other open-graded friction courses. No comprehensive program had been undertaken in South Carolina before this study to examine the extent and severity of stripping.

Selected reviews of literature on stripping [3,4], reviews of many of the recent tests [5–7], and more than 120 abstracts of articles obtained from the Transportation Research Board's Highway Research Information Service corroborate the pervasiveness of stripping. One recommendation made at a Federal Highway Administration (FHWA) Workshop on Stripping held in Atlanta, GA, was to undertake a survey to identify and quantify the extent of asphalt stripping in flexible pavements.

In June 1983, the South Carolina Department of Highways and Public Transportation (SCDHPT) initiated a comprehensive research program in cooperation with the Department of Civil Engineering at Clemson University. The project included the following tasks: (1) determine the extent and severity of stripping in South Carolina; (2) identify the major causes of stripping in South Carolina highways; and (3) develop effective measures to eliminate or minimize stripping. The third phase of the study, completed in June 1985, involved identifying laboratory tests that seem most effective in identifying measures to prevent or to minimize stripping.

Procedure

Extensive Coring

Programs of extensive and intensive coring were initiated to obtain pavement cores from throughout the state to provide baseline data on the extent and severity of stripping. Highways to be sampled for the extensive coring program were selected by SCDHPT personnel to encompass a wide range of pavement and material types. During the extensive program of coring, two cores were obtained every 3.2 km (2 miles) by use of a water-cooled, truck-mounted core drill. Specimens were cored from the outer wheel path, 61 cm (2 ft.) from the pave-

TABLE 1—*Locations of pavement coring.*

Highway	Traffic Group	Location	km	Miles	Cores
US 76	J	Marion to Mullins to Nichols	25.7	16	23
US 301	J	Dillon to Latta	11.3	7	15
US 17 Bypass	J	Myrtle Beach (N and S of US 501)	30.6	19	70[a]
US 17	K	North and South of Georgetown	67.6	42	94[a]
US 17	K	North of Charleston	25.7	16	34
US 17	L	South of Charleston	12.9	8	16
I-95	O	Jasper County	22.5	14	29
I-26	O	N of SC 33 to Cola, Orangebg County	51.5	32	64
US 321	J	Chester to Winnsboro	35.4	22	75[a]
US 276	N	Rest Area N of I-26 to Simponsville	33.8	21	62[a]
I-20	M	Bishopville to Kershaw County Line	54.7	34	71
US 276	J	Union to Whitmire	22.5	14	58[a]
US 78	M	Montmorenci to Denmark	54.7	34	98[a]
US 378	L	SC 391 to Lexington to I-26	41.8	26	53
SC 9	K	Lancaster east toward Pageland	22.5	14	58[a]
I-85	O	SC 153 to Georgia State Line	45.1	28	56
SC 11	E	US 25 to US 276	19.3	12	24
US 176	J	Ballentine to Pomaria	29.0	18	41
US 521	J	Lancaster to Kershaw	24.1	15	30
US 25	L	SC 183 to North Carolina State Line	37.0	23	44
I-20	M	West of Columbia	53.1	33	96[a]
US 501	J	Latta to Marion	19.3	12	23
SC 28	J	Abbeville S to McCormick Cty. ln	19.3	12	58[a]
US 123	J	US 25 (G'ville to SC 93 (Clemson)	41.8	26	82[a]
US 17L	K	G'town Cty to 2-miles past McCllnv'le	3.2	2	50[a]
		TOTALS	804.6	500	1324

[a]Intensive coring.

ment edge stripe. Locations for coring within each 3.2-km (2-miles) subsection of the highway were identified by use of random numbers. Preliminary assessment of the stripping of cores was conducted, when possible, at the site, and characteristic features of the pavement and site were recorded on prepared forms. Characteristics of the site that were recorded included a visual assessment of the pavement condition (for example, sound or flushed, extent and type of distress or cracking; cut, fill, or level section; condition of drainage; and presence of standing water). Pavement coring locations are listed in Table 1 and included approximately 806 km (500 miles) and 1324 cores.

Pavements were cored even where no surface evidence of stripping, such as flushing or raveling, was apparent. Pavement specimen cores from both traffic directions were obtained; however, minor relocation of coring sites was sometimes necessary to maintain safety during sampling. Each core was placed in a plastic bag, identified by a label, and stored on its top, flat surface in compartmentalized wooden crates to minimize handling stresses.

After pavement cores were brought back to the laboratory at Clemson, they were weighed, measured, photographed, and sawed into construction layers. Testing of cores from the extensive sampling program involved measuring the compacted specific gravity of each specimen using vacuum saturation (51-cm

TABLE 2—*Arrays of* C *and* F.

Values of C	Values of F
C = coarse aggregate stripping	F = fine aggregate stripping
1 = less than 10%	1 = less than 10%
2 = 10 to 40%	2 = 10 to 25%
3 = more than 40%	3 = more than 25%

(20 in.) mercury for 5 min) and immersion for 24 h at 60°C (140°F) before testing in indirect tension (deformation rate = 5.1 cm/min [2 in./min]) at 25°C (77°F) using a recording Marshall machine. Conditioning procedures were needed to provide identical initial conditions for determining specific gravity. A visual estimate of the stripping evident in each specimen cross section obtained after indirect tension testing was obtained using the tentative rating procedure developed by the Georgia Department of Transportation (DOT).

The Georgia DOT stripping rating S is calculated by assigning values of C and F in the expression of $S = (C + F)/2$, where C and F are shown in Table 2.

The advantage of the Georgia DOT procedure for rating stripping is its simplicity in its separate rating of stripping in the coarse aggregate and the fine aggregate fractions. The procedure, however, requires training for consistent interpretation of results. The low correlation coefficients for operator error evaluation shown in Table 3 indicate that, during an evaluation of 168 specimen cross sections, Operator C, who received less instruction, was a relatively inaccurate rater of stripping damage.

The visual strip ratings obtained from the extensive coring program were used in assessing the extent and severity of asphalt stripping in South Carolina pavements. The extensive sampling program was unbiased, and no attempt was made to obtain core specimens from only distressed areas. Sampling provided specimens that were representative of pavements selected for testing by SCDHPT personnel.

Intensive Coring

Intensive coring was completed for selected sites for which more thorough testing of as-received and dried core specimens was needed to characterize

TABLE 3—*Correlation coefficients for operator error evaluation* (n *=168).*

Operator			
A	B	C	D
1.0000	0.7397	0.4525	0.8468
0.7397	1.0000	0.5671	0.8298
0.4525	0.5671	1.0000	0.5108
0.8468	0.8298	0.5108	1.0000

specimen strength, extent of stripping, asphalt content, and percent air voids. Intensive pavement coring was completed for selected sites by obtaining 16 cores from the outer wheel path (WP) and 16 cores from the middle of the outside traffic lane (ML).

In those locations where intensive coring was used, 14 cores were obtained using the water-cooled, truck-mounted drill. In addition, two cores from the outer wheel path and two cores from the mid-lane were obtained using a dry coring procedure. In this procedure, a portable core drill (Fig. 1*a*) was used to obtain undisturbed 100 mm- (4-in.) diameter core specimens. Carbon dioxide and compressed air were used to cool the core drill bit and to blow the aggregate dust from the hole. During extensive coring with the water-cooled drill, it was observed that the drill coolant water eroded the specimen and, in some instances, obscured the extent and severity of stripping. Use of dry coring (which was slower than the wet coring procedure) enabled project personnel to detect in-situ pavement moisture that could not be detected (Fig. 1*b*) using the water-cooled drill.

Core specimens from the intensive sampling program were weighed, measured, photographed, and sawed into construction layers as were the cores obtained from the extensive coring program. Some cores were tested in indirect tension after being allowed to dry. Other cores were saturated to approximately 55 to 80% saturation in accordance with the procedure described by Tunnicliff and Root [*8*]. This procedure involved saturating the cores using a vacuum of 63.5 cm (25 in.) of mercury applied for 5 min. This was followed by a 24-h submersion period in distilled water at 60°C (140°F). The layers were cooled to 25°C (77°F) in water before testing in indirect tension. Visual stripping ratings were made using tentative Georgia DOT procedures described earlier. Rice specific gravities were obtained only for specimens obtained from the intensive coring program.

Photographs of each specimen were taken, and the locations of each core site were plotted on a county map. The location of each core site was precise so that identification of materials and ages of pavement layers could be established from information on file in the principal SCDHPT office in Columbia. Other data measured at the site where intensive cores were taken were similar to those recorded for the extensive coring program.

Experimental Results

Extent and Severity of Stripping

Stripping was found to be pervasive in South Carolina and no area of the state was free of some amount of stripping. The extent and severity of stripping were characterized for each highway by the summaries of the frequency of severely stripped layers as noted from visual estimates of specimens in indirect tension. Specimens whose visual strip ratings were 2.5 or 3 were classified as severely stripped. The mean visual strip rating and the standard deviation for each high-

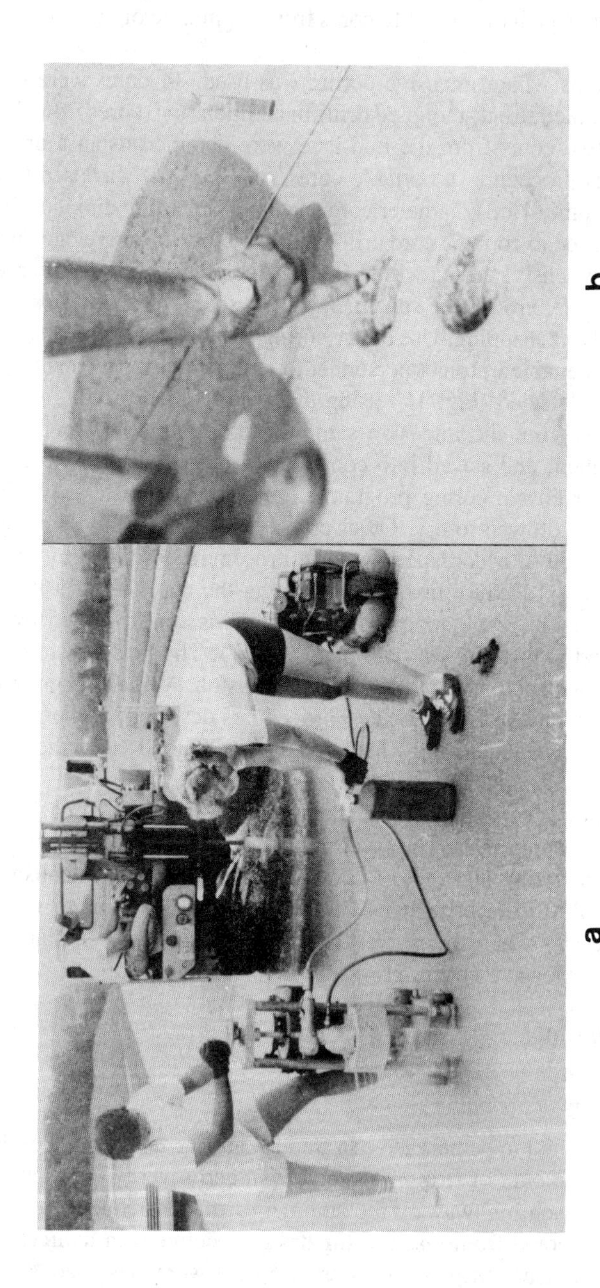

FIG. 1—(a) *Air-cooled pavement core drill and* (b) *in-situ moisture evident from dry coring.*

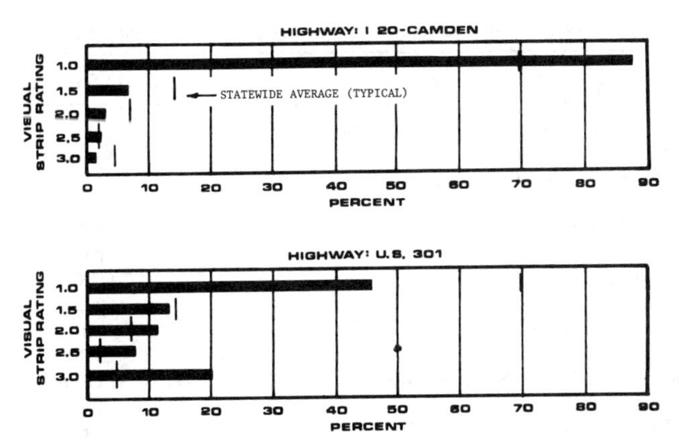

FIG. 2—*Typical strip rating histograms for two South Carolina highways.*

way sampled were recorded. Indirect tensile strengths were measured but were not strongly correlated with visual stripping (coefficient of correlation: $r = 0.58$) although there was strong correlation for some mixtures. Strength of different mixtures composed of different aggregates and asphalt contents vary widely as Kennedy noted [7].

A frequency of severe stripping (that is, visual strip ratings 2.5 and 3) of 8.1% was measured for the entire program of statewide sampling. Strip ratings of individual highways when compared to the state average (8.1%) served as measures of severity. For example, results for two of the highways (US 301 and I-20-Camden) are depicted in Fig. 2, which shows good performance on I-20 and more severe stripping on US 301. For US 301, visual stripping levels of 2.5 and 3.0 accounted for 28% of all pavement layers examined while I-20 had only 5% of the layers with levels of 2.5 and 3.0. Each of the 26 highway sections cored in the extensive sampling program was similarly characterized.

Stripping Distress Related to Aggregate Source

Visual evidence of stripping was related to the source of aggregate as noted from SCDHPT records. In most instances, a single quarry supplied the coarse aggregate and most of the other granular material; however, in a few instances, blended aggregates from two quarries were used. Stripping frequencies (expressed as a percentage) ranged from 34 to 0%. Those aggregate sources for which stripping frequencies exceeded the state average (8.1%) and those for which stripping frequency was less than half the state average are listed in Table 4. Those quarries for which the stripping frequency was high should receive priority attention for assessing the effectiveness of anti-strip agents. Some of the aggregates from quarries for which stripping frequency was less than half the state average have been required, based on past stripping distress, to use anti-stripping agents.

TABLE 4—*Relationship between aggregate source and stripping frequency.*

Quarry Location	Percent Stripped
QUARRIES FOR WHICH STRIPPING EXCEEDED STATE AVERAGE	
Augusta, GA	34.0
Dreyfuss, North Columbia, SC	24.1
Palmetto, Columbia, SC	17.3
Jefferson, SC	12.5
Bennettsville, SC	11.8
Cash, SC	11.5
Blair, SC	10.5
Georgetown, SC	9.1
Liberty, SC	8.2
QUARRIES FOR WHICH STRIPPING WAS LESS THAN HALF THAT OF STATE AVERAGE	
Pacolet, SC	1.1
Lakeside, Greenville, SC	2.6
Cayce, SC	4.1
Anderson, SC	0.0
Hendersonville, NC	0.0

Stripping Distress Related to Mixture Type

The frequency of severe stripping and its relation to various mixture types was computed from results of the extensive coring and testing program. It is generally recognized that mixtures have different potentials for stripping when different sources of aggregate are used. Data for all mixtures sorted by quarries were obtained for this project and the averages are listed in Table 5.

Data show that stripping was slight in sand asphalt mixtures (generally used in base courses) and in hot-mixed asphalt bases. Binder T-2 was more susceptible to strip or weaken in locations where water could come in contact with the mixture. The most severe stripping was observed in Surface T-3 mixtures. This mixture has been widely used below plant-mixed seal courses (PMSC), a South Carolina open-graded friction course. PMSC layers were not severely stripped; however, stripping of PMSC mixes tended to occur at the interface of the PMSC and the supporting layer. PMSC layers were generally less than ten years old.

Plant-mixed seal courses (PMSC) and sand asphalt layers (generally used as base courses in aggregate scarce locations in the coastal plain province) had lower than average stripping rates. The PMSC layers had thick coatings of asphalt, which may improve resistance to moisture damage. Sand asphalt mixtures were well-coated with asphalt and did not exhibit extensive stripping damage. Information regarding asphalt source and type used in all mixes was not readily available, and consequently no correlation of stripping with these factors could be made.

The average indirect tensile strength of severely stripped specimens was 6.96 kg/cm^2 (99.0 lb/in.^{2}) and was significantly different (a = 0.05) from the average indirect tensile strength (12.02 kg/cm^2, 171.0 psi) of all other specimens. Tensile strength varies considerably with mixture type. In the strength analysis noted here, strengths of sand asphalt mixtures and PMSC layers were not included because they were not primary structural mixes.

TABLE 5—*Stripping distress related to mixture type.*

Mixture Type	Total Number of Layers	Percent Stripped
PMSC	352	4.83
Surface T-1	754	7.69
Surface T-1A	107	13.08[a]
Surface T-2	656	8.07
Surface T-3	822	14.35[a]
Binder T-1	135	6.66
Binder T-2	656	9.75[a]
Binder T-3	238	0.84
Binder T-4	108	3.70
Base	398	1.26
Sand asphalt	743	2.55

[a]Mixture exceeds state average strip rating (8.1%).

Some mixtures placed directly over portland cement concrete pavement exhibited stripping by emulsification of the asphalt. These mixes were located on slabs that retained water that seeped in from overlying pavement layers.

Stripping Under Open-Graded Friction Courses

It was hypothesized that the use of open-graded friction courses or PMSC increased the frequency of stripping. Field data collected from this study enabled this hypothesis to be examined quantitatively. A search of all PMSC surface layers was initiated to identify those mixtures that were used immediately below the PMSC layers. A summary of results from this search is shown in Table 6. The statewide average stripping frequency under open-graded friction courses was 18.7% compared with a statewide average of 8.1% for all pavement layers. In general, PMSC layers and, in many instances, the layer immediately below the PMSC were newer layers because the use of PMSC mixes is approximately ten years old in South Carolina. Surface Type T-3 was more widely used below PMSC than other surface types. This mixture exhibited a stripping frequency of 22.2%–approximately 2½ times the state average.

Stripping Related to Pavement Age

Stripping of pavement layers was related to pavement age. Specimens from the extensive coring program were grouped into five-year age intervals. Within each interval the percentage of pavement layers that were severely stripped was computed. Data are shown in Fig. 3.

The highest percentage (13.2%) of severely stripped pavement specimens falls in the six- to ten-year interval. The use of open-graded friction courses in South Carolina likewise originated approximately ten years ago. It is possible that moisture-related damage became more widespread as a result of using the open-graded friction course. Pavements that are older than 20 years range up to a maximum of 28 years.

TABLE 6—*Stripping immediately below plant mixed seal course (PMSC).*

Highway	Mix Type Under PMSC	Percent Stripped	(Fraction)
US 17-GN	Surface T-3	22.2	(6/27)
US 17-GS	Surface T-1	0.0	(0/4)
US 17-CN	Surface T-3	37.5	(6/16)
US 76	Surface T-2	20.0	(2.10)
US 301	Surface T-3	50.0	(7/14)
US 176-B	Surface T-3	80.0	(20.25)
I 26	Surface T-3	15.6	(10/64)
I 95	Surface T-3	6.9	(2/29)
US 25	Surface T-1	0.0	(0/4)
US 78	Surface T-1	10.0	(2/20)
	Surface T-3	13.3	(2/15)
US 378	Surface T-3	0.0	(0/10)
US 321	Surface T-1	9.1	(3/33)
US 123	Surface T-1	0.0	(0/4)
I 85	Surface T-3	0.0	(0/32)
	Surface T-1	0.0	(0/24)
I-20 CAMDEN	Surface T-1	100.0	(4/4)
US 17-LI	Surface T-2	0.0	(0/2)
	Surface T-3	12.5	(2/16)
		AVERAGE 18.7	(66/353) overall

SUMMARY FOR LAYERS IMMEDIATELY BELOW PMSC

SURFACE T-1 (9/93) = 9.7% stripped
T-2 (2/12) = 16.7% stripped
T-3 (55/248) = 22.2% stripped

Relationship Between Stripping and Traffic Group

Stripping was pervasive and unpredictable based on traffic group. A summary of stripping within each traffic group (E through O had increasing truck counts) was analyzed and showed (Tablc 7) that there was no obvious correlation between truck traffic and stripping. Other factors, such as presence of paved shoulders, recent use of anti-stripping agents in some interstate pavements, and so forth, would have to be known to make significant statistical statements.

Effect of Type of Section

Data from the field survey were analyzed to determine whether the type of section (that is, level, cut, or fill section) from which the core was taken had any effect on the severity of stripping. The stripping frequency was calculated for each mix type for each section (Table 8). In addition, statistical tests of differences in indirect tensile strengths were used to identify whether the type of section had a significant effect ($a = 0.05$) on the extent and severity of stripping. Overall, 10.2% of the layers from level sections were stripped, whereas 7.6 and 4.6% of layers from cut and fill sections, respectively, were stripped.

Sand asphalt layers and base courses did not exhibit much stripping regardless of the section. PMSC layers from level sections had a stripping frequency of 10% while the same mixes exhibited 1.1 and 2.9% stripping frequencies in cut and fill sections, respectively.

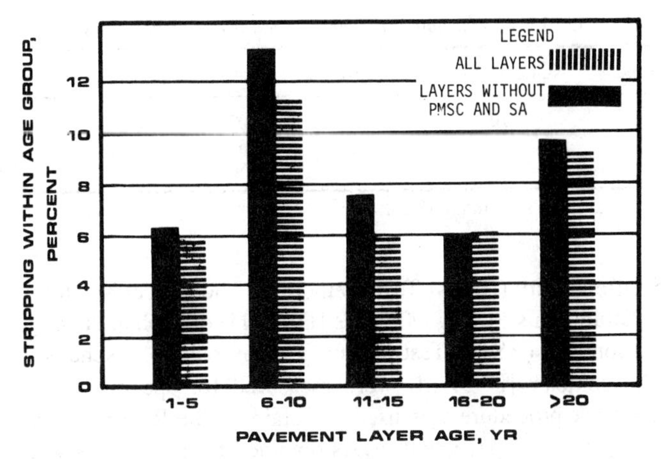

FIG. 3—*Stripping percent related to pavement layer age.*

Saturation and Air Voids in Pavement Layers (Intensive Coring)

Specimens obtained from the intensive sampling program tested using saturation and testing procedures proposed by Tunnicliff and Root [*8*] could not always be saturated to 55 to 80% saturation. Overall, approximately 67% of the specimens were saturated to that level. Other specimens were less than 55% saturated. Some specimens had low air voids and these were difficult to saturate even when the vacuum was applied for up to 30 min. Other specimens that were cracked internally may have been saturated to the desired 55 to 80% level; however, these specimens lost water from the crack when the specimen was withdrawn from the water and then dried while obtaining saturated surface-dry weights.

TABLE 7—*Summary of percent stripping for traffic groups.*

	Traffic Group (Increasing Truck Count)							
Mix Type	E	I	J	K	L	M	N	O
PMSC	···	···	19.8[a]	0.0	0.0	2.6	···	0.0
Surface T-1	···	1.7	9.0	14.9[a]	2.0	12.6[a]	3.0	7.0
Surface T-2	···	19.0[a]	5.2	22.2[a]	···	···	···	0.0
Surface T-3	0.0	···	19.0[a]	18.0[a]	0.0	20.9[a]	···	9.5[a]
Surface T-4	···	···	···	11.5[a]	···	···	···	···
Binder T-1	···	···	2.5	···	···	···	12.7[a]	···
Binder T-2	0.0	17.2[a]	13.2[a]	5.0	5.8	21.8[a]	···	4.6
Binder T-3	···	···	···	···	···	0.5	···	···
Binder T-4	···	···	···	···	···	···	···	3.8
Surface T-1-A	···	···	···	···	···	21.9[a]	···	0.0
Base	···	···	···	···	1.1	···	···	1.3
Sand asphalt	···	···	9.4[a]	0.0	10.0[a]	0.0	···	···
Total	0.0	10.6[a]	12.2[a]	8.9[a]	2.8	7.4	7.0	3.4

[a]Exceeds state average strip rating (8.1%).

TABLE 8—*Stripping frequencies for various section types.*

Section Type	Number of Layers	Number of Layers Stripped	Percent Stripped
Level	1053	107	10.2[a]
Cut	1384	63	4.6
Fill	2066	156	7.6

[a]Exceeds state average strip rating (8.1).

Rice specific gravity (ASTM Test Method for Theoretical Maximum Specific Gravity of Bituminous Paving Mixtures [D 2041]) of asphalt concrete mixtures and asphalt contents (ASTM Test Method for Viscosity of Asphalts by Vacuum Capillary Viscometer [D 2172]) were measured for specimens. The supplementary dry back procedure was used in obtaining the Rice maximum specific gravity. Altogether 235 pavement layers obtained from the cores were tested in this way. Experimental determination of maximum specific gravity and asphalt content enabled air-void contents to be measured. In addition, it was possible, when original job-mix formulas were available, to calculate the theoretical quantity of air voids in pavement layers.

Open-graded friction courses and sand asphalt mixtures typically have high air-void contents. The latter mixes are generally constructed using local sands in the lower coastal plain areas. The natural sand mixes are not well-graded, and generally have low Marshall stabilities, high voids in the mineral aggregate (VMA), and high total air-void contents.

Of the 235 specimens tested in the intensive coring program, 61% had air-void contents that exceeded 6%. When all sand asphalt layers and plant-mixed seal courses (high air-void mixtures) were removed from the specimen population, 51% of the remaining 189 pavement layers tested had an air-void content that exceeded 6%. These high air-void contents (Fig. 4) and low densities may have resulted from earlier South Carolina method-based compaction specifications. Many of these mixtures were T-3 surface mixtures for which an above-average stripping frequency was noted. It is generally recognized that high air-void contents can be a source of stripping through water retention, especially when the mixture is immediately below open-graded friction courses. Retained indirect tensile strength of wheel path and mid-lane specimens conditioned using the Tunnicliff and Root saturation procedure [*8*] were 72.5 and 86.8%, respectively.

Retained Marshall stability is currently used by SCDHPT to evaluate the need for anti-stripping agents. One group of specimens is tested after 30- to 60-min immersion in 60°C (140°F) water while a second group of specimens is tested after 24-h immersion. The minimum retained Marshall stability (that is, wet strength/dry strength) formerly required by the state of South Carolina was 75%; however, it has recently been increased to 85%. Tests are made on specimens made with job-mix formula aggregates containing the optimum asphalt content based on 50-blow Marshall compaction.

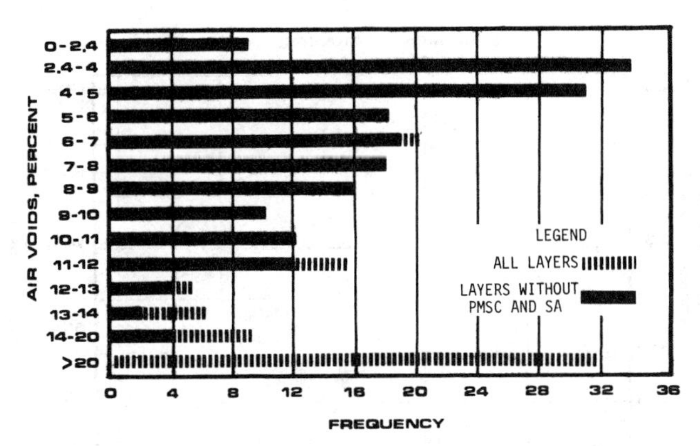

FIG. 4—*Air-voids histogram.*

Many of the core specimens were observed after testing to contain cracks with an accumulation of fine granular material in the crack. These granular materials had, in some instances, accumulated from upward movement of granular base materials into the overlying mixture. This results in a loss of support and is similar to pumping action that may occur in rigid pavements.

Specimen Strength Related to Pavement Layer Thickness

Indirect tensile strengths were measured for all pavement layers obtained from the extensive coring program. Thin surface layers and open-graded friction courses were less than 25.4-mm (1-in.) thick after being trimmed using a masonry saw. In addition, sand asphalt pavement layers, constructed with natural coastal plain sands, possessed low stability and low indirect tensile strength. A summary of the indirect tensile strength of all pavement layers sampled and of layers without PMSC and sand asphalt layers indicated that indirect tensile strength, on the average, was a function of specimen thickness. This corroborates the Georgia DOT practice that indirect tensile strength should not be measured for specimens less than 25.4-mm (1-in.) thick. Results are shown in Fig. 5.

Conclusions

The conclusions presented here are based on the results of laboratory and field studies discussed previously. Approximately 8% of the South Carolina pavement specimens from the outer wheel path showed serious visual evidence of stripping. Dry coring can be used to detect in-situ moisture that might go undetected using water-cooled core drills.

Stripping is pervasive to some extent throughout the South Carolina highways selected for this inventory. Aggregate sources that have been associated with high frequency of stripping have been identified and selected so that cost-effective treatments for minimizing stripping damage can be developed. Some mixture

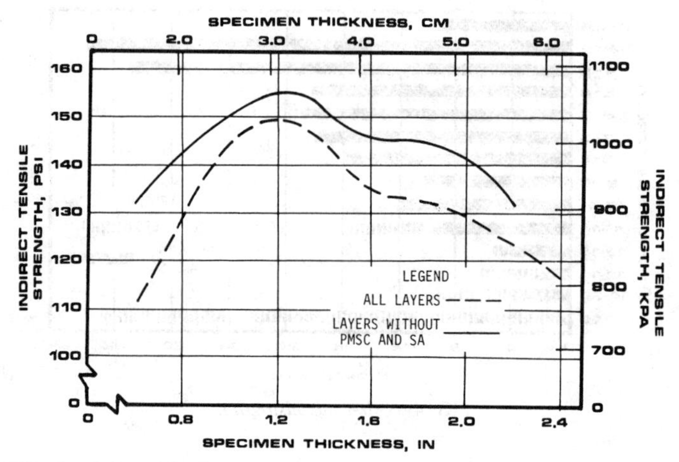

FIG. 5—*Relationship between indirect tensile strength and specimen thickness.*

types are susceptible to stripping; however, it is not clear whether these mixtures are inherently susceptible or whether insufficient compaction and concommitant high air voids caused problems [*9*]. More than half of all dense-graded specimens tested had more than 7% air voids.

The frequency of stripping was more severe and frequent under open-graded friction courses than elsewhere. Related to this was an above average frequency of stripping observed in pavements in the six- to ten-year age group. No relationship between traffic group and stripping was measured; however, level sections, especially those with plant-mixed seal courses had more stripping than cut or fill sections.

Indirect tensile strength appears to be a useful test to measure wet and dry strengths to assess susceptibility of field or laboratory specimens to stripping for specimens at least 25.4-mm (1-in.) thick.

Acknowledgments

Messrs. Chip Voyles, James Benson, Howe Crockett, and Ms. Krista Martini assisted in obtaining field and laboratory data. Messrs. Milford Wald and Milton Lore assisted with machining and technical support. Numerous SCDHPT maintenance personnel helped the authors collect data safely under sometimes heavy traffic. Dr. Hoke Hill assisted with statistical analysis of data. The research study described herein was supported by the Federal Highway Administration and the South Carolina Department of Highways and Public Transportation. The contents of this paper reflect the views of the authors who are solely responsible for the facts and the accuracy of the data presented. The contents do not necessarily reflect the official views of either the Federal Highway Administration or the South Carolina Department of Highways and Public Transportation.

References

[1] Transportation Research Board, *Special Report 202,* America's Highways—Accelerating the Search for Innovation, Washington, DC, 1984, pp. 71–72.

[2] Stapler, T., "A Survey of Bituminous Pavement Distress Attributable to Water Damage as Reported by the SASHTO States," a *survey* conducted by the Vice-chairman, AASHTO Region 2, Operating Subcommittee of Materials, 1981, 10 pp.

[3] Majidzadeh, K. and Brovold, F. N., Effect of Water on Bitumen-Aggregate Mixtures, *Highway Research Board Special Report 98,* Washington, DC, 1968.

[4] Taylor, M. A. and Khosla, N. P., Stripping of Asphalt Pavements; State of the Art, 62nd Annual Meeting of the Transportation Research Board, Washington, DC, January 1983.

[5] Lottman, R. P., Laboratory Test Method for Predicting Moisture-Induced Damage to Asphalt Concrete, *Transportation Research Record 842,* Transportation Research Board, Washington, DC, 1982, pp. 88–95.

[6] Gilmore, D. W., Lottman, R. P., and Scherocman, J. A., "Use of Indirect Tension Measurements to Examine the Effect of Additives on Asphalt Concrete Durability," presented at the Annual Meeting of the Association of Asphalt Paving Technologists, Phoenix, AZ, April 1984, 12 pp.

[7] Kennedy, T. W., Roberts, F. L., and Lee, K. W., "Evaluation of Moisture Susceptibility of Asphalt Mixtures Using the Texas Freeze-Thaw Pedestal Test," *Proceedings of the Association of Asphalt Paving Technologists,* Vol. 51, 1982, pp. 327–241.

[8] Tunnicliff, D. G. and Root, R. E., "Testing Asphalt Concrete for Effectiveness of Antistripping Agents," *Proceedings of the Association of Asphalt Paving Technologists,* Vol. 52, 1983, pp. 535–560.

[9] Kennedy, T. W., Roberts, F. L., and McGennis, R. B., "Effects of Compaction Temperature and Effort on the Engineering Properties of Asphalt Concrete Mixtures," *Placement and Compaction of Asphalt Mixtures, ASTM STP 829,* F. T. Wagner, Ed., American Society for Testing and Materials, 1984, pp. 48–66.

Hossein Takallou,[1] *R. Gary Hicks,*[1] *and James E. Wilson*[2]

Evaluation of Stripping Problems in Oregon

REFERENCE: Takallou, H., Hicks, R. G., and Wilson, J. E., **"Evaluation of Stripping Problems in Oregon,"** *Evaluation and Prevention of Water Damage to Asphalt Pavement Materials, ASTM STP 899,* B. E. Ruth, Ed., American Society for Testing and Materials, Philadelphia, 1985, pp. 22–48.

ABSTRACT: Before 1974, stripping in hot mix asphalt concrete was considered to be a relatively minor problem in Oregon. During the three-year period from 1974 to 1977, several problems were noted during and after construction of asphalt pavements. One of the major problems associated with asphalt pavements was its initial reduced resistance to raveling. This problem developed after the oil embargo in 1974. In recent years, problems with stripping and associated pavement deterioration have continued such that in 1983 amine-type antistrip agents were incorporated into over 20% of all projects constructed.

The state of Oregon's specification (before 1984) attempts to ensure good performance of asphalt mixtures against stripping. For example, the determining factor as to the suitability of a particular mixture against stripping has been American Association of State Highway and Transportation Officials (AASHTO) T-165, index of retained strength (IRS). If the IRS of a mix at the design asphalt content is below 70% with the asphalt and aggregate to be used in the paving, an antistrip agent is used to satisfy this requirement. A 70% minimum resilient modulus ratio after vacuum saturation and freeze-thaw conditioning is used to determine the suitability of a particular mixture against stripping.

This paper addresses the current problem of stripping in the state of Oregon. Field studies conducted in 1983 were designed to identify causes of recent raveling problems in central Oregon (high elevation and severe climate areas). The investigations involved field condition surveys and coring, reviewing project construction records, mix designs, and asphalt concrete test data (construction cores and compaction control test results), and numerous tests on cores obtained from the roadway.

From the information collected, several factors were found to be contributing to the observed raveling and recommendations made for correcting these deficiencies have been included in the 1984 specifications.

KEY WORDS: asphalts, pavements, stripping, asphalt stripping, pavement life, resilient modulus, antistrip agents

[1]Research assistant and professor of civil engineering, respectively, Oregon State University, Department of Civil Engineering, Corvallis, OR 97331.

[2]Assistant engineer of materials, Oregon Department of Transportation, Highway Division, 2950 E. State St., Salem, OR 97310.

In recent years, problems with stripping and the associated pavement deterioration have troubled highway engineers in the state of Oregon. Because of this increasing awareness of asphaltic pavement failures caused by stripping of asphalt cement from the aggregate, the Oregon Highway Division, required prior to 1984, an increased use of amine-type antistripping additives. In 1984, the use of lime-treated aggregate in asphalt concrete paving to prevent asphalt stripping was initiated.

This paper addresses specifically the problem of stripping in the state of Oregon. The history of water-related problems, types of antistripping agents used, and results of field studies conducted in 1983 on Oregon pavements are discussed.

History of Stripping Problems

Before 1974 stripping in hot mix asphalt concrete was considered to be a relatively minor problem. During the three-year period from 1974 to 1977, several problems were noted throughout the Pacific Northwest during and after construction of asphalt pavements. For example, in Oregon, construction and short-term performance problems that were seldom experienced before 1974 developed, including low mix cohesion [*1*]. This resulted in either raveling or early surface deterioration caused by stripping and increased the need for antistripping agents.

In an effort to resolve these problems, the Oregon State Highway Division developed a questionnaire, distributed in 1974 and 1976 to Regional and Project Construction Engineers, to establish the extent and causes of pavement problems experienced in 1974 through 1976. The results of both questionnaires indicated there was an increased need for the use of antistripping additives to satisfy the specified 70% minimum index of retained strength (IRS) using American Association of State Highway and Transportation Officials (AASHTO) Effect of Water on Cohesion of Compacted Bituminous Mixtures (T-165). This indicated that asphalts or aggregates supplied to Oregon users in 1974 through 1976 were more subject to stripping or water sensitive than those supplied earlier.

In 1982, another study undertaken by the Oregon Department of Transportation for the Federal Highway Administration (FHWA) was carried out [*2*]. The purpose of this study was to evaluate the effect of material sources, void content, and additive type on the index of retained strength or retained resilient modulus after freeze-thaw conditioning. Furthermore, the study objective was to identify and quantify the extent of asphalt stripping in flexible pavements. To satisfy the purpose of the study, a total of 20 projects were evaluated. The results of this study clearly indicated that (1) present mix design procedures may not always detect problems from asphalt-aggregate stripping; (2) aggregate quality and asphalt sources appear to relate to low values for IRS and modulus ratio; (3) significant differences existed for IRS and modulus ratio values for construction mix design specimens, submitted mix specimens, and laboratory batched specimens; (4) level of compaction greatly affected the compressive strength, however, the IRS values show little change; (5) freeze-thaw conditioning greatly affected modulus and modulus ratios; and (6) the use of additives generally increases both the modulus ratio and IRS [2].

Specification Requirements (Before 1983)

The state of Oregon's specifications attempt to ensure good performance of asphalt mixtures against stripping. For example, the determining factor as to the suitability of a particular mixture against stripping is IRS. From 1958 to 1983, if the IRS values for a design mix were below 70%, treatment of the asphalt with an amine-type additive would be required. Since 1980, the Oregon State Highway Division also has used resilient modulus ratios for unconditioned and freeze-thaw conditioned specimen as a factor to determine the suitability of a particular mixture to provide resistance against stripping. If the modulus ratio is below 70% (at the recommended asphalt content), an antistrip agent is required.

Table 1 summarizes the average properties of antistrip agents, which have been evaluated by the Oregon Department of Transportation (ODOT) in recent years. Most of the properties indicate the workability of the agents under construction and mixing conditions. The most important properties of antistripping agents are viscosity and heat stability. Viscosity is a measure of the ability of the agent to blend uniformly with asphalt and to coat the aggregate adequately. Heat stability of the antistripping agent is a measure of the ability of the agent to maintain adhesion properties after storage in hot asphalt (140 to 180°C) and during paving plant mixing. In most specifications, the actual values for each property are not specified, and the suitability of each agent is a function only of its ability to prevent stripping under standard test conditions such as (AASHTO T-165) or the modulus ratio test.

Field Study, 1983

In 1983, the Oregon Department of Transportation incorporated antistripping agents into asphalts for asphalt concrete mix designs in over 20% of their projects (Fig. 1). Despite this, there are still major stripping and raveling problems, particularly in mountainous areas (elevations over 760 m). In a field survey conducted during Feb. 1983, several projects showed considerable problems. This investigation involved field condition surveys and coring, and an analysis of laboratory tests on cores obtained from the roadway. In response to the initial survey and these problems, a second survey was conducted in June 1983. The discussions which follow describe both of these surveys.

Projects Evaluated in Feb. 1983

On 21–25 Feb. 1983, a total of 14 projects (Table 2) were surveyed by Oregon DOT personnel. All projects were paved between the Summers of 1978 and 1980 and are in areas of moderate to heavy freeze-thaw and snowfall (elevations > 760 m).

TABLE 1—*Selected antistrip agents evaluated by Oregon before 1982.*

Property	Brand A	Brand B	Brand C	Brand D	Brand E	Brand F	Brand G	Brand H
Flash (COC)°C	138	149	224	121	210	185	107	135
% of Total Distillate to								
190°C	0	0	0	0	...	...	2.7	...
225°C	1.5	1.8	0	1.5	...	...	6.3	...
260°C	6.7	10.5	0	3.1	...	...	22.3	...
316°C	62.7	64.9	8.9	36.2	...	...	69.6	...
Residue From Distillation to 360°C	33.0	43.0	77.5	35.0	...	...	44	...
% water	0	0	0	0	...	...	0	...
Specific gravity	1.018	1.020	0.945	1.005	0.954	0.970	0.972	0.966
Weight, lb/gal[a]	8.48	8.50	7.87	8.37	7.95	...	...	...
ph	11.29	12.44	10.94	10.98	9.032	9.58	...	...
Kinematic viscosity, original								
38°C, cs[b]	193	414	255	467	253	112	...	...
60°C, cs	50	92	72	156	75	31	...	...
Kinematic viscosity, residue after RTFC oven aging								
38°C, cs	did not flow	...	53	3324	...	...	...	...
60°C, cs	355	166	6	46	...	...	...	...

[a]1 cs = 10^{-3} Pa · s.
[b]lb/gal = 119.8 kg/m^3.

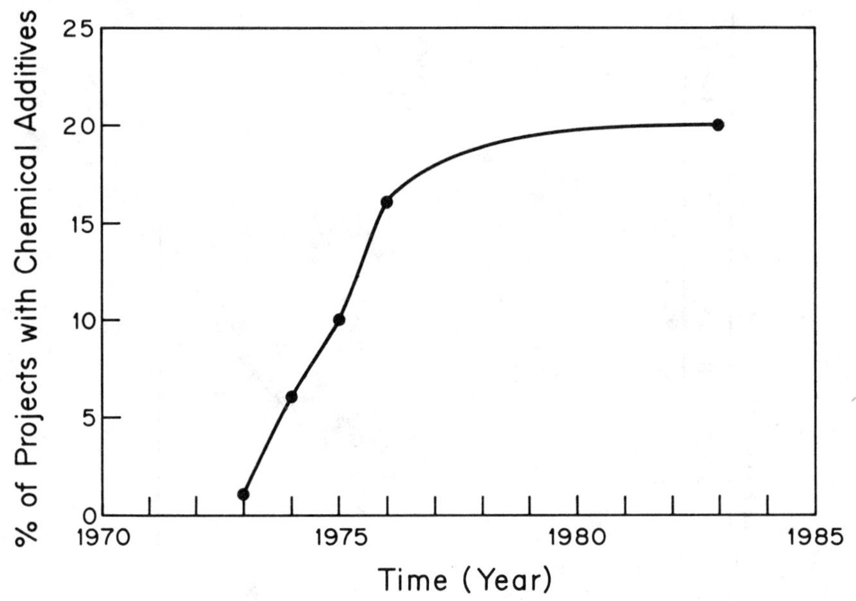

FIG. 1—*Use of antistripping additives in Oregon (1973 through 1983).*

Of the projects for which construction mix designs were available (Table 3), four exhibited tendencies for stripping when tested without additive treatment of asphalt as indicated by the IRS values of less than 70% at the recommended asphalt contents. These included (1) Willamette Junction-Chemult, (2) Collier State Park-Williamson River, (3) Ski Bowl Frontage Road-Summitt Meadow Road, and (4) Warm Spring Junction-Salmon Road Curves. The February inspection showed all four of the above mentioned projects to be in either poor or very poor condition, despite the fact they all exhibited passing IRS values with chemical additives.

Table 4 presents the results of tests on submitted cores from five of these projects taken in Feb. 1983. All projects had high IRS values (>80%) and did not require additives. These included (1) Hackett Drive-Crescent, (2) Lava Butte Lookout-Sugar Pine Butte Road, (3) Odell Lake-Camp Mukulla, (4) Glen Baker-Lava Butte, and (5) Bend S. C. L.-Murphy Road. Of the five projects for which submitted cores were tested, all had high modulus values before water saturation or freeze-thaw conditioning; all had lower modulus values after conditioning. As indicated, four of the five pavements inspected were in poor condition. The Bend S. C. L.-Murphy Road project was the only one in good condition.

From the information and data collected, the following factors appeared to be possible contributors to the distress:

1. The air voids in the wearing and base courses of several projects were high.
2. There were large variations in aggregate gradations and asphalt contents when compared with normal mix tolerances (Table 5).

TABLE 2—*Oregon pavements inspected by ODOT in Feb. 1983.*

Project Number	Project	Highway	Year Completed	Traffic ADT[a]	Snowfall	Pavement Condition	Additive Used
1	Zigzag River—Ski Bowl Frontage Road	U.S. 26	1980	3 400	heavy	poor	Brand C
2	Ski Bowl Frontage Road—Summit Meadow Road	U.S. 26	1978	3 400	heavy	very poor	Brand G
3	Warm Springs Junction—Willow Creek	U.S. 26	1978	2 400	heavy	poor	none
4	Fremont Junction—Hacket Drive	U.S. 97	1980	3 300	heavy	good	none
5	Madras/Prineville Hwy.—Culver Hwy. Sec.	U.S. 97	1979	5 000	light	good	none
6	Willamette Junction—Chemult	U.S. 97	1980	3 900	heavy	poor	Brand D
7	Lava Butte Lookout—Sugar Pine Butte Road	U.S. 97	1980	8 200	heavy	poor	none
8	Hackett Drive—Crescent	U.S. 97	1981	3 200	heavy	poor	none
9	Fuego Road—Forge Road	U.S. 97	1980	3 500	heavy	poor	Brand D
10	Odell Lake—Camp Makualla Road	State 18	1979	2 600	heavy	poor	none
11	Collier State Park—Williamson River	U.S. 97	1978	3 600	heavy	very poor	Brand D
12	Bend S.C.L.—Murphy Road	U.S. 97	1980	21 700	moderate	good	none
13	Warm Springs Int—Salmon River Curves	U.S. 26	1978	2 800	heavy	poor	Brand G
14	Glen Baker—Lava Butte	U.S. 97	1977	9 600	moderate	poor	none

[a]Average daily traffic.

TABLE 3—*Mix design results, Feb. 1983 survey.*

Project	Type of Mix	Asphalt Type	Recommended Asphalt Content, %		Index of Retained Strength				Comments
					w/o Additive		w/Additive		
			Surface	Base	Surface	Base	Surface	Base	
(1) Zigzag River—Ski Bowl Frontage Rd	"B" AC	AR4000[a]	5.0	5.5	86	93	100	100	Brand C antistrip additive used. Above test results indicate additive not required to obtain 70% IRS specifications.
(2) Ski Bowl Frontage Rd—Summit Meadow Rd	"B" AC	AR4000[a]	5.0	5.5	31	36	82	85	Recommend use of 0.4% Brand G additives for base and 0.5% Brand G additive for surface to satisfy 70% IRS specifications.
(3) Warm Springs Jct—Willow Creek	"B" AC	AR4000[a]	5.3	5.8	81	86	...	...	Recommend minimum 4.8% asphalt content for 70% IRS specifications.
(4) Fremont Jct—Hackett Dr	"B" AC	AR4000[b]	5.5	6.0	76	83	...	...	Additives are not required.
(5) Madras/Prineville Hwy—Culver Hwy Sec	"B" AC	AR4000[b]	4.7	5.2	87	91	...	...	Additives are not required.
(6) Willamette Jct—Chemult	"B" AC	AR4000[a]	5.9	6.5	0	0	85	87	Recommend use of AR4000 treated with 0.5% Brand D liquid antistrip additives to satisfy 70% IRS specifications.
(7) Lava Butte Lookout—Sugar Pine Butte Rd	"B" AC	AR4000[b]	5.4	6.0	91	95	...	...	Test results indicate antistrip additive not required to obtain 70% IRS specification.
(8) Hackett Drive—Crescent	"B" AC	AR2000 & AR4000[a]	5.9	6.5	91	95	...	...	It is recommended that AR2000 grade rather than AR4000 asphalt be used to provide best resistance to pavement damage from raveling, thermal cracking, and fatigue cracking.

(9) Fuego Road—Forge Road	"B" AC	AR4000[a]	5.8	6.3	74	80	90	97	Recommend use of 0.2% Brand D antistrip additive to ensure 70% IRS specification.
(10) Odell Lake—Camp Makualla Road	"B" AC	AR4000[b]	5.4	6.0	80	90	···	···	Test results indicate antistrip additive not required to obtain 70% IRS specification.
(11) Collier State Park—Williamson RV	"C" AC	AR4000[a]	7.4	6.5	0	0	100	86	Recommend use of AR4000 treated with Brand D liquid antistrip additives to satisfy 70% IRS specification.
(12) Bend S.C.L.—Murphy Rd	"B" AC	AR4000[b]	5.4	6.0	91	95	···	···	Additives are not required.
(13) Warm Springs Int—Salmon R. Curves	"B" AC	AR4000[a]	5.0	5.5	24	40	82	85	Minimum asphalt content for 70% IRS specification is base 5% AC with 0.4% Brand G additive base 4.5% AC with Brand G additive wearing.
(14) Glen Baker—Lava Butte	"E" AC	AR4000[b]	6.5	7.0	98	100	···	···	Test results indicate antistrip additive not required to obtain 70% IRS specification. Based on comparison of IRS mix design tests using AR4000 and AR8000 asphalt and low temperature properties of the asphalt, it is recommended AR4000 be used rather than AR8000.

[a]California Valley.
[b]California Valley and Coastal Blend.

TABLE 4—*Oregon DOT core evaluation, Feb. 1983.*

Project	Hackett Drive-Crescent		Lava Butte Lookout-Sugar Pine Butte Road		Odell Lake-Camp Makualla		Glen Baker Road-Lava Butte		Bend S.C.L. Murphy Road	
	Average	Range	Average	Range	Average	Range	Average	Range	Average	Range
Pavement condition	poor		poor		poor		poor		good	
Gradation, % passing										
25 mm (1 in.)	100	100 to 100	100	100 to 100	100	100 to 100	...	...	100	100 to 100
20 mm (3/4 in.)	98	90 to 100	99	100 to 99	99	100 to 97	100	100 to 100	99	97 to 100
13 mm (1/2 in.)	87	80 to 92	85	83 to 89	87	86 to 90	99	100 to 90	83	81 to 87
10 mm (3/8 in.)	75	64 to 84	72	60 to 78	73	67 to 77	89	78 to 93	71	67 to 78
6 mm (1/4 in.)	60	46 to 69	58	54 to 63	55	50 to 58	62	51 to 70	56	52 to 63
2 mm (No. 10)	33	26 to 39	33	28 to 35	27	25 to 30	25	16 to 32	31	28 to 35
0.425 mm (No. 40)	16	14 to 18	16	14 to 18	12	11 to 13	13	10 to 17	16	14 to 18
0.074 mm (No. 200)	5.4	4.7 to 5.9	6.6	5.3 to 10.3	6.0	5.6 to 6.6	5.3	4.0 to 6.7	5.8	4.7 to 6.7
Asphalt content, %	5.6	4.8 to 6.6	5.7	5.1 to 6.2	5.9	5.3 to 6.6	6.3	4.6 to 8.0	5.4	5.2 to 5.8
Coating, %	83	75 to 95	81	55 to 95	85	75 to 95	46	40 to 50	88	93 to 83
Air voids	6.12	2.4 to 11	4.6	3.4 to 6.7	10.4	5.6 to 13.3	11.4	8.6 to 14.8	6.5	4.4 to 9.6
Modulus, $\times 10^3$ psi at 25°C										
(a) unconditioned	999	550 to 1522	1030	549 to 1302	869	673 to 1052	936	844 to 1342	1340	1093 to 1622
(b) vacuum saturated	747	429 to 1071	814	457 to 1004	577	415 to 700	776	590 to 988	1119	914 to 1354
(c) freeze-thaw	573	394 to 748	729	375 to 983	501	344 to 642	731	598 to 957	803	676 to 1014
Modulus ratio										
b/a, %	76	66 to 92	80	74 to 88	66	60 to 73	72	64 to 75	83	80 to 86
c/a, %	62	38 to 81	71	66 to 77	57	51 to 67	68	54 to 83	60	56 to 63
Asphalt properties										
pen at 25°C (77°F), dmm	60	43 to 88	43	39 to 56	28	24 to 35	52	21 to 163	45	38 to 57
viscosity at 135°C (275°F), cs	253	195 to 279	543	399 to 619	716	592 to 1046	578	220 to 892	621	604 to 695
viscosity at 60°C (140°F), poises	1909	1016 to 2574	3534	2295 to 4564	7670	5098 to 10252	9923	2315 to 18550	4668	2686 to 6385

NOTE: 1 in. = 2.54 cm, 1 psi = 6.895 kPa, 1 centipoise = 10^{-3} Pa · s, 1 poise = 10^{-1} Pa · s.

TABLE 5—*Acceptable tolerances—Oregon specifications.*

Item	Tolerance
Sieve Size	
20 mm (¾ in.)	±6
13 mm (½ in.)	±6
6 mm (¼ in.)	±6
2 mm (No. 10)	±4
0.425 mm (No. 40)	±4
0.074 mm (No. 200)	±2
Asphalt content	±0.5

3. All projects were in areas of moderate to heavy snowfall.
4. The high IRS values determined during mix design had little effect on resulting performance.
5. The average modulus ratios (c/a) for all projects were approximately 70% or less.

Projects Evaluated in June 1983

In June 1983, a total of 15 projects were surveyed in central Oregon by Oregon State University, Chevron USA, and Oregon DOT personnel while 6 projects were surveyed by Washington DOT. Projects evaluated in these surveys were paved between 1972 and 1981. Nearly all of the projects were in a harsh climate area with moderate to heavy traffic. Eleven of these projects were the same as those inspected in Feb. 1983. Project descriptions and locations are given in Table 6 and Fig. 2.

Most of the projects evaluated were overlays ranging in thickness from 38.1 to 63.5 mm (1.5 to 2.5 in.). The types of failures observed consisted of surface raveling or strip raveling or both. Surface raveling was initiated with a loss of fine aggregate from the surface of the pavement (Fig. 3) or some loss of larger material (Fig. 4), which eventually led to severe surface raveling (Fig. 5). This raveling was definitely accelerated through the use of studded tires or chains on pavement surface with high void contents or both (Fig. 6).

The strip raveling generally showed up as a narrow, deeply raveled section longitudinal to the roadway. The strip-raveled sections observed were usually 101.6 to 203.2 mm (4 to 8 in.) in width, a few feet to several hundred feet in length, and with a depth not greater than the wearing surface overlay (Fig. 7). There were not any consistent distress patterns observed. For example, the strip raveling did not always occur in the wheel paths, as would be expected. Frequently, the strips were closer together than a wheel path width and were located at the center and outer edge of the traffic lane (Fig. 8).

For each project surveyed, samples of the aggregate used were obtained for testing. Both 101.6- and 152.4-mm (4- and 6-in.) diameter cores were taken across the panel (Figs. 9 and 10) in an attempt to isolate the cause of the strip

TABLE 6—*Oregon pavements inspected in June 1983.*

Project Number	Project	Highway	Time of Year Completed	Traffic ADT[a]	Snowfall	Pavement Condition	Type of Plant	Mix Temperature °F	Contractor
1[b]	Zigzag River—Ski Bowl Frontage Road	U.S. 26	10/80	3400	heavy	poor	drum	300	Babler
2[b]	Ski Bowl Frontage Road—Summit Meadow Rd	U.S. 26	10/78	3400	heavy	very poor	batch	325	Babler
3	Warm Springs Junction—Salmon River Curves	U.S. 26	10/78	2800	heavy	poor	batch	325	Babler
4	Salmon River Curves—Frog Lk (Federal Job)	U.S. 26	11/78	4200	heavy	excellent	batch	...	Babler
5	Warm Springs Junction—Willow Creek	U.S. 26	9/78	2400	heavy	good	drum	...	Stinson
6	Simnasho Road—Jefferson County Line	U.S. 26	1972	2650	heavy	good	batch	...	Coats
7	Madras/Prineville Hwy—Culver Hwy Sec.	U.S. 97	9/79	5000	light	excellent	batch	300	Coats
8	Forest Boundary—Kaiwia Spgs (Federal Job)	Cascade Lakes	6/82	1000	moderate	good	drum	...	Babler
9[b]	Plainview Road—Deschutes River	U.S. 20	7/80	5700	moderate	excellent	batch	295	Coats
10[b]	Lava Butte Lookout—Sugar Pine Butte Road	U.S. 97	6/80	8200	heavy	poor	batch	350	Coats
11	Fremont Junction—Hackett Drive	U.S. 97	8/80	3300	heavy	good	batch	305	Coats
12	Hackett Drive—Crescent	U.S. 97	7/81	3200	heavy	fair	drum	225	Babler
13[b]	Willamette Junction—Chemult	U.S. 97	8/80	3900	heavy	poor	drum	275	Babler
14	Fuego Road—Forge Road	U.S. 97	10/80	3500	heavy	fair	batch	280	Stukel-Rock
15[b]	Odell Lake—Camp Makualla Road	State 58	9/79	2600	heavy	poor	drum	270	Compton

NOTE: $T_c = (T_F - 32)/1.8$.
[a]Average daily traffic.
[b]Also inspected by Washington Department of Transportation

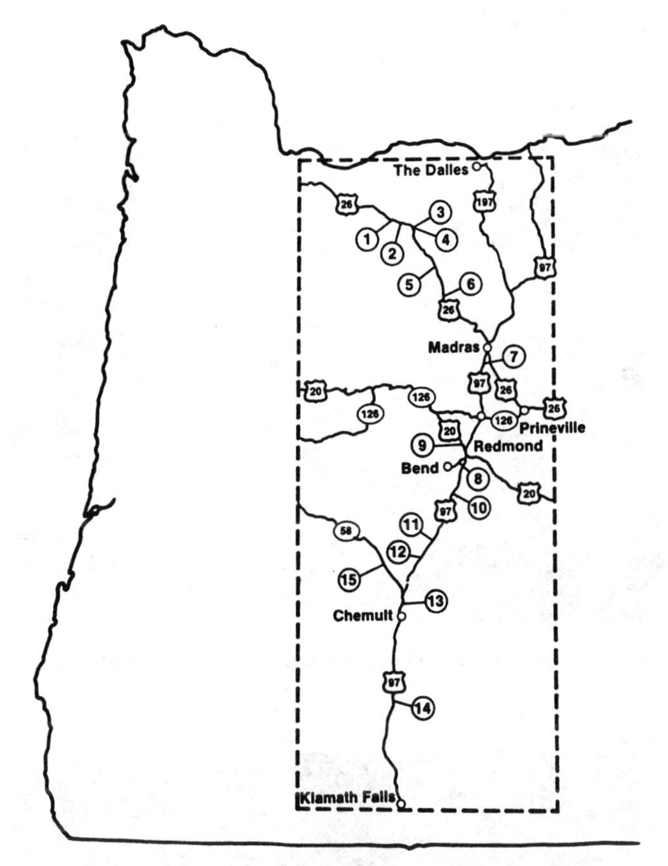

FIG. 2—*Project locations in central Oregon, June 1983 survey.*

raveling. Table 7 summarizes the aggregate source used in the specified projects. Properties of aggregates are listed in Table 8. In the following sections, studies performed by Chevron Research Corporation [*3*] and for the Oregon DOT [*4–7*] on the identified projects are discussed.

Chevron Research Corporation

The projects identified in Table 6 were also studied by Chevron Research Corporation [*3*]. For each project, cores were obtained (a total of 168) and the following determined:

(1) void content,
(2) asphalt content,
(3) aggregate gradation, and
(4) asphalt properties.

FIG. 3—*Surface raveling (Forest boundary-Kaiwia Springs, FHWA job).*

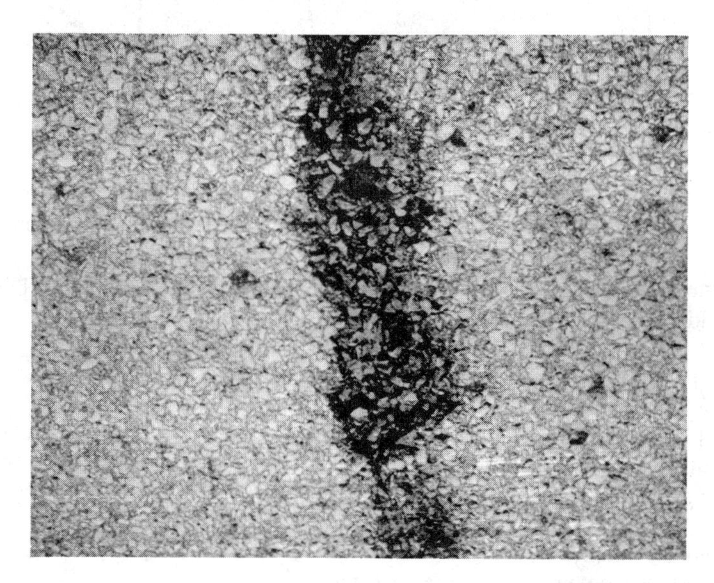

FIG. 4—*Loss of surface coarse aggregate and asphalt (Warm Springs Junction-Salmon River Curves).*

FIG. 5—*Surfacing raveling throughout travel lane (Lava Butte-Sugar Pine Butte Road).*

FIG. 6—*Potholes throughout the lane (Ski Bowl Frontage Road-Summit Meadow Road).*

FIG. 7—*Strip raveling (Salmon River Curves-Frog Lake, ODOT versus FHWA).*

FIG. 8—*Strip raveling at middle of driving lane (Lava Butte-Sugar Pine Butte Road).*

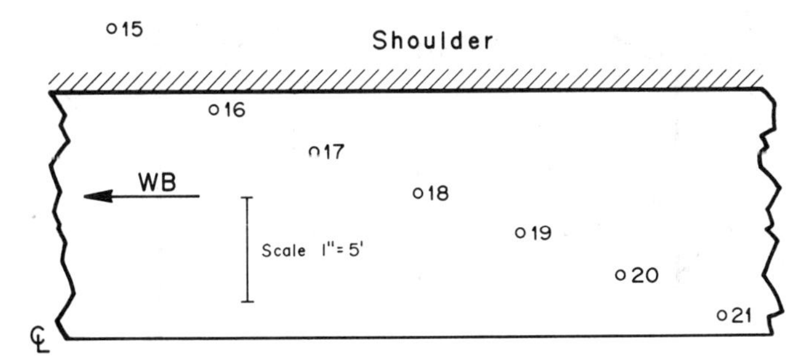

FIG. 9—*Core layout for Lava Butte Lookout-Sugar Pine project.*

In addition, diametral modulus values, both as-received (ASTM Method for Indirect Tension Test for Resilient Modulus of Bituminous Mixtures [D 4123]) and after moisture conditioning [*9*] were tested on all projects.

The results of their test program indicated:

1. The aggregates used on most projects had relatively high water absorption values ranging from near 0 to 6% (Tables 7 and 8).
2. High air voids existed in many of the projects.
3. The retained modulus (b/a) on the projects ranged from 30 to 110%. In general, the retained modulus values were lower for courses with high void contents.
4. Considerable variation in appearance and distress existed along the lane.
5. The recovered properties of the asphalt appeared reasonable, considering the grades and sources used as well as the age and condition (void content) of the pavement.
6. Mix moisture, from either external or internal sources, appears to have contributed to the early distress. This is also confirmed in the work reported by Kim et al [*8*].

ODOT Studies

At the request of the Oregon Department of Transportation, six projects were evaluated by the Washington DOT [*4*] while two projects were evaluated by Oregon State University [*7*] to determine the cause of the observed raveling. The projects evaluated by WSDOT included:

(1) Zigzag River-Ski Bowl Frontage Road,
(2) Ski Bowl Frontage Road-Summit Meadow Road,
(3) Plainview Road-Deschutes River,
(4) Lava Butte Lookout-Sugar Pine Butte Road,
(5) Willamette Junction-Chemult, and
(6) Odell Lake-Camp Makualla Road.

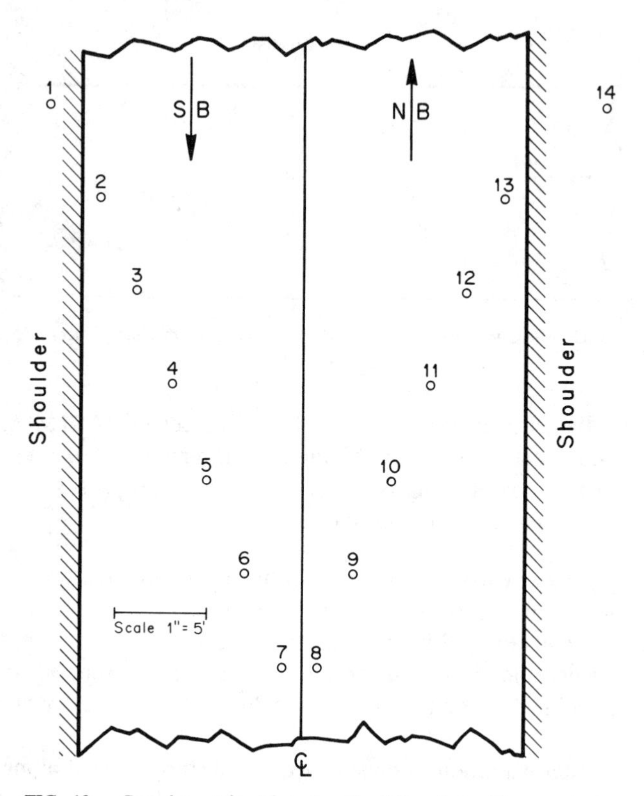

FIG. 10—*Core layout for Plainview Road-Deschutes River project.*

Oregon State University evaluated Projects 3 and 4 above. All projects were previously described in Table 6.

Table 9 summarizes the results of tests on cores by the Washington DOT [*4*]. The premature failure of five of the six projects reviewed were categorized into two groups: (1) general raveling and (2) strip raveling or pothole raveling, which is not necessarily associated with the wheel path tracking of the pavement, or both. In all five failing projects strip raveling was observed. Only three projects showed surface raveling and strip raveling in combination. These included (1) Zigzag River-Ski Bowl Frontage Road, (2) Ski Bowl Frontage Road-Summit Meadow Road, and (3) Lava Butte Lookout-Sugar Pine Butte Road. The results of the evaluations identified the following factors as contributing to the observed problems:

1. All projects were located in severe climatic conditions, including moisture and freezing.
2. Asphalt contents of the projects appear to be low based on Washington's experience.

TABLE 7—*Aggregate source and properties for projects evaluated, June 1983—Source.*

Project Number	Project	Source Name
1	Zigzag River—Ski Bowl Frontage Road	Quarry Excavation of Truck Ramp at MP 50 Mt Hood Hwy
2, 3, 4	Ski Bowl Frontage Road—Summit Meadow Road, Warm Springs Junction—Salmon River Curves, and, Salmon River Curves—Frog Lk (Federal Job)	Trillium Quarry
5	Warm Springs Junction—Willow Creek	Quarry—SE ¼ Sec 10 T55 R9E
6	Simnasho Road—Jefferson County Line	Quarry—MP 116.90
7	Madras/Prineville Hwy—Culver Hwy Sec.	Coats Madras Quarry
8	Forest Boundary—Kaiwia Springs (Federal Job)	35-09-002
9, 10, 11	Plainview Road—Deschutes River, Lava Butte Lookout—Sugar Pine Butte Road, and Fremont Junction—Hackett Drive	Gravel Cake Pit
12	Hackett Drive—Crescent	Russell Pit Gravel
13, 15	Willamette Junction—Chemult, and Odell Lake—Camp Makualla Road	Waker Mt. Quarry
14	Fuego Road—Forge Road	Stukle Mt.—Tulane Farms Quarry

3. The air voids in most projects were high. This is a reflection of the low percentage of asphalt or the compaction of the mat or both.
4. Aggregate gradation was out of specification on a number of projects.
5. There were instances of excessive heating of the mix caused by long haul distances.
6. Some pavements were constructed late in the year. This would not allow the pneumatic tire action of traffic to seal the pavement surface on warm days and allow moisture to become trapped in the mix.
7. Strip raveling was related to incomplete pickup of the mix from windrows or excessive mix moisture or both.

A total of 21 cores were shipped to Oregon State University from the Plainview Road-Deschutes River project (excellent condition) and the Lava Butte Lookout-Sugar Pine Butte Road project (poor condition) for evaluation. Cores 1 through 14 were taken from the Dalles-California Highway (Lava Butte Lookout project), and Cores 15 through 21 from the McKenzie Bend Highway (Plainview Road-Deschutes River project). The core layout for both projects is shown in Figs. 9 and 10. These two projects were both constructed by the same contractor, with the same asphalt cement and aggregate. The Plainview Road project, however, contained an antistrip agent while the Lava Butte Lookout project did not. Further, the Plainview Road project had a relatively short haul distance (<8 km) [<5 miles] and is located in drier climatic area, while the Lava Butte Project had a haul distance of 32 to 48 km (20 to 30 miles) and is located in a climatic area that gets heavy snow in the winter.

TABLE 8—*Aggregate source and properties for projects evaluated, June 1983—Property.*

Aggregate Source	Bulk Specific Gravity	Apparent Specific Gravity	Absorption, %
Quarry excavation of track Ramp at MP50 Mt. Hood Hwy			
coarse	2.72	2.79	0.9
fine	2.67	2.77	1.29
Trillium Quarry			
coarse	...	2.79[a]	...
fine	...	2.79[a]	...
Quarry SE ¼ Sec 10 T55 R9E			
coarse	...	2.74[a]	...
fine	...	2.74[a]	...
Quarry MP 116.9			
coarse	2.76	2.82	0.06
fine	2.60	2.71	0.11
Coats Madras Quarry			
coarse	2.81	2.95	1.72
fine	2.75	2.97	2.67
35-09-002			
coarse	...	2.79[a]	...
fine	...	2.85[a]	...
Gravel Cake Pit			
coarse	...	2.78[a]	...
fine	...	2.76[a]	...
Russell Pit Gravel			
coarse	2.53	2.71	2.71
fine	2.49	2.72	3.41
Waker Mt Quarry			
coarse	2.61	2.75	1.98
fine	2.52	2.77	3.69
Stukle Mtn.—Tulane Farms Quarry			
coarse	2.51	2.72	3.10
fine	2.32	2.72	6.36

[a]Measured with AASHTO T-133.
[b]Coarse 20-6 mm (¾ to ¼ in.) sieve. Fine 6 mm-0 (¼ to 0 in.) sieve.

Table 10 shows the results of the density and void analyses of the cores taken across the panel. As indicated, the average voids for both projects are in the acceptable range with little variation across the panel.

Table 11 summarizes the results of resilient modulus testing on the samples at the following conditions:

1. Unconditioned—The cores were tested at room temperature (20°C ± 1).
2. Conditioned (tested at 16°C ± 2)—The cores were vacuum saturated for 2 h, frozen for 15 h at −17°C, thawed in a water bath for 24 h at 60°C, then conditioned in water for 3 h at 25°C before testing [*9*]. The modulus was corrected from the test temperature of 16 to 20°C using temperature correction factors developed by the Asphalt Institute [*10*].

TABLE 9—*Core evaluation by WSDOT, June 1983.*

Project	Zigzag River Ski Bowl Frontage Rd.		Ski Bowl Frontage Rd. Summit Meadow Rd.		Plainview Rd. Deschutes River		Tolerances
	Average	Range	Average	Range	Average	Range	
Pavement condition	very poor		very poor		excellent		. . .
Gradation, % passing							
25 mm (1 in.)	100	100	100	100	100	100	±6
20 mm (¾ in.)	100	100	99.6	98 to 100	99.9	100	±6
13 mm (½ in.)	89.5	85 to 93	88.8	84 to 94	88.3	87 to 91	±6
10 mm (⅜ in.)	. . .	. . .	. . .	. . .	. . .	. . .	±6
6 mm (¼ in.)	59.4	54 to 67	57.3	52.69	61.3	54 to 68	±6
2 mm (No. 10)	32.4	27 to 41	26.1	23 to 32	29.2	27 to 36	±4
0.425 mm (No. 40)	16.9	15 to 20	11.9	11 to 14	12.6	11 to 16	±4
0.074 mm (No. 200)	6.6	5.6 to 7.4	3.80	3.3 to 4.9	6.1	5.4 to 7.5	±2
Asphalt content, %	5.17	4.66 to 6.3	5.23	4.8 to 5.6	5.5	5.4 to 6.0	±0.5
Air voids, %	10.4	. . .	10.0	9.1 to 12.9	8.1	. . .	
Modulus, 10^3 psi							
(a) unconditioned	. . .	. . .	501	238 to 634	. . .	. . .	
(b) vac. sat.	. . .	. . .	408	151 to 570	. . .	. . .	
(c) freeze-thaw	. . .	. . .	217	65 to 404	. . .	. . .	
Modulus ratio, %							
b/a	. . .	. . .	79	63 to 88	. . .	. . .	
c/a	. . .	. . .	41	12 to 70	. . .	. . .	
Asphalt properties							
pen at 25°C, dmm	56.3	42 to 76	53.9	52 to 56	90.8	80 to 97	
vis at 135°C, cs	289.7	245 to 361	385.5	375 to 396	517	. . .	
vis at 60°C, poises	3283	2009 to 4299	3851.8	3776 to 4040	3398.2	3037 to 3777	

TABLE 9—*Continued.*

	Lava Butte Lockout Sugar Pine Butte Rd.		Willmette Jct. Chemult		Odell Lake Camp Mokuall Rd.		
Project	Average	Range	Average	Range	Average	Range	Tolerances
Pavement condition	poor		poor		poor		
Gradation, % passing							
25 mm (1 in.)	100	100	100	100	100	100	±6
20 mm (¾ in.)	99.2	97 to 100	99.7	97 to 100	98.9	96 to 100	±6
13 mm (½ in.)	84.4	81 to 97	88.6	83 to 94	85.8	78 to 93	±6
10 mm (⅜ in.)	...	...	...	...	...	...	±6
6 mm (¼ in.)	58.7	56 to 65	59.6	49 to 70	55.8	44 to 62	±6
2 mm (No. 10)	33.9	31 to 35	27.2	20 to 36	25.8	21 to 29	±4
0.425 mm (No. 40)	14.0	12 to 16	11.3	8 to 14	11.5	8 to 12	±4
0.074 mm (No. 200)	5.8	4.6 to 7.8	6.2	4.1 to 8.1	5.6	4.3 to 6.6	±2
Asphalt content, %	5.7	5.5 to 6.0	5.7	5.0 to 6.2	5.8	5.7 to 6.0	±0.5
Air voids, %	7.3	...	8.4	...	9.1	...	
Modulus, 10^3 psi							
(a) unconditioned	1083	965 to 1302	...	...	849	779 to 914	
(b) vac. sat.	843.5	817 to 967	...	...	572	482 to 655	
(c) freeze-thaw	764.5	663 to 983	...	...	483	445 to 548	
Modulus ratio, %							
b/a	78	74 to 88	...	...	67	62 to 72	
c/a	71	66 to 76	...	...	57	55 to 60	
Asphalt properties							
Pen at 25°C, dmm	77.8	54 to 91	53.2	48 to 76	81.2	71.94	
Vis at 135°C, cs	...	...	340.0	326 to 354	...	...	
Vis at 60°C, poises	3823.3	3389 to 4293	3779.1	2076 to 4427	2793.2	2412 to 3451	

NOTES: 1 in. = 2.54 cm, 1 psi = 6.895 kPa, 1 cs = 10^{-3} Pa · s, 1 poise = 10^{-1} Pa · s.

TABLE 10—*Density and voids analysis of cores from Oregon pavements, June 1983, Oregon State University study.*

	(a) Lava Butte Lookout—Sugar Pine Butte Road							
Core ID	1	2	3	4	5	6	7	average
Specific gravities measured	2.329	2.342	2.388	2.383	...	2.393	2.284	2.353
Maximum theoretical specific gravity	2.488	...	...	...	...	...	...	2.488
Air voids, % from measured specific gravity	6.39	5.87	4.02	4.22	...	3.82	8.12	5.41
	(b) Lava Butte Lookout—Sugar Pine Butte Road							
Core ID	8	9	10	11	12	13	14	average
Specific gravities measured	2.410	2.432	2.429	2.433	2.433	2.447	2.415	2.428
Maximum theoretical specific gravity	2.488	...	...	...	...	...	...	2.488
Air voids, % from measured specific gravity	3.14	2.25	2.37	2.21	2.21	1.61	2.93	2.39
	(c) Plainview Road—Deschutes River							
Core ID	15	16	17	18	19	20	21	average
Specific gravities measured	2.315	2.386	2.359	2.331	2.345	2.345	...	2.012
Maximum theoretical specific gravity	2.487	...	...	...	...	...	...	2.487
Air voids, % from measured specific gravity	6.92	4.06	5.15	6.27	5.70	5.70	...	5.63

TABLE 11—*Modulus properties of cores from Oregon pavements, June 1983, Oregon State University.*

			(a) LAVA BUTTE LOOKOUT—SUGAR PINE BUTTE ROAD					
Core ID	1	2	3	4	5	6	7	average
Unconditioned								
(a) M_R, (20°C ± 1) psi	603 000	483 000	332 000	367 000	...	...	597 000	476 400
Conditioned								
(c) M_R, (16°C ± 2) psi	699 000	236 000	519 000	524 000	...	511 000	677 000	527 666
corrected M_R[a]	660 000	227 000	480 000	486 000	...	476 000	645 000	495 000
Retained strength, %								
c/a	1.16	0.49	1.56	1.43	...	...	1.13	1.15
c/a (corrected)	1.09	0.47	1.45	1.32	...	...	1.08	0.90
			(b) LAVA BUTTE LOOKOUT—SUGAR PINE BUTTE ROAD					
Core ID	8	9	10	11	12	13	14	average
Unconditioned								
(a) M_R, (20°C ± 1) psi	339 000	286 000	233 000	275 000	268 000	290 000	330 000	288 714
Conditioned								
(c) M_R, (16°C ± 2) psi	506 000	435 000	409 000	435 000	492 000	486 000	587 000	478 571
corrected M_R[a]	474 000	410 000	375 000	410 000	403 000	390 000	502 000	423 000
Retained strength, %								
c/a	1.49	1.52	1.76	1.58	1.68	1.78	1.67	1.64
c/a corrected M_R[a]	1.40	1.43	1.61	1.49	1.50	1.52	1.47	1.49
			(c) PLAINVIEW ROAD—DESCHUTES RIVER					
Core ID	15	16	17	18	19	20	21	average
Unconditioned								
(a) M_R, (20°C ± 1) psi	429 000	358 000	378 000	464 000	404 000	526 000	460 000	431 286
Conditioned								
(c) M_R, (20°C ± 1) psi	380 000	383 000	147 000	294 000	312 000	207 000	404 000	253 857
Retained strength, %								
c/a	0.89	1.10	0.39	0.63	0.85	0.39	0.88	0.73

NOTE: 1 psi = 6.895 kPa.
[a]Modulus corrected to 20°C [*10*].

Note that although the average retained modulus (c/a) for all projects was above 0.70, there was considerable variation across the panel. The results of this evaluation indicate:

1. For the two projects studied, there was considerable variation in c/a ratio across the panel. This is probably caused by variations in asphalt content.
2. The air voids of both projects were in an acceptable range, with little variation across the panel.
3. The project in excellent condition had an antistrip agent (Brand H) while the one in poor condition did not.

Discussion of Results

The results of the Feb. and June 1983 surveys, as well as the results of the FHWA [*3*] and HP&R studies [*8*], have led to some significant changes in asphalt paving specification in recent years. Some of these include the following:

1. Treatment of asphalt concrete aggregates with 1% lime. The required mellowing of treated aggregates for a minimum of one (lime slurry) or five (dry hydrated lime) days at a moist condition before use in a paving mix. Lime treatment of aggregates is required on a mandatory, or when required, basis on projects with more than 3000 tons of paving. Projects in areas over 4023-m (2500-ft) elevation, freeze-thaw conditions, or with aggregate quality problems require mandatory lime treatment of aggregates. Nonmandatory projects require lime treatment when the IRS of the mix without treatment is less than 80% at the design asphalt content. Generally, the IRS is near 100% after treatment [*2*].
2. The design asphalt contents of the paving mixes have been increased slightly by reducing the design voids from 4–7% to 3–5%. This is expected to reduce raveling by increasing the IRS or modulus ratios.
3. The minimum IRS value has been increased from 70 to 75%. However, if the IRS is less than 80%, lime is required. This could result in the use of an asphalt additive if the lime alone is not able to increase the IRS above 75%.
4. Reduced mix moisture from 1.0 to 0.7% at the plant. The Chevron study [*3*] indicated internal moisture is directly related to the occurrence of the strip raveling. The HP&R study [*8*] also pointed out that excess moisture effects both modulus and fatigue life. By reducing this moisture, the chance for raveling is expected to decrease considerably.
5. Because of the high voids on several of the projects, emphasis has been placed on their reduction. This should come about because of higher asphalt contents, increasing the minimum air temperature for compaction, and requiring pneumatic rollers on all projects.
6. To reduce the amount of variation in gradation and asphalt content across the panel, better control on the use of storage silos (for drum mixers) and on pickup machines during laydown operations have been instituted. For storage silos, batchers or rotating chutes have been installed to reduce segregation.

A summary of all changes in Oregon specifications from 1974 to 1984 is given in Table 12.

Conclusions and Recommendations

The results of this paper indicate the following:

1. In 1983, the Oregon Department of Transportation incorporated chemical antistripping agents into the mix design for 20% of their total projects.
2. Because of recent problems, lime treatment of aggregates is now required for high elevation projects or problem pavement areas in Oregon's 1984 specification.
3. The Chevron and Oregon DOT studies both indicate that many of the projects had high air voids (>10%). Further, the Oregon studies showed considerable variation in aggregate gradation and asphalt content across the panel.
4. The analysis of laboratory tests on cores obtained from the roadway indicate that the average retained resilient modulus is greater than 70%; however, there is considerable variation across the panel.
5. The observations made in June 1983 indicated that projects constructed with the same materials and by the same contractor performed differently. The one with the antistrip agent performed well while the other did not.
6. The asphalts are prone to stripping when exposed to severe climatic conditions. Stripping appears to be reduced by controlling moisture in mix, film thickness, and construction techniques that minimize segregation. The most promising factor, however, appears to be the use of lime or additives that change the characteristics of the asphalt.

The results of recent studies led the ODOT to make the following changes for the 1984 season:

(1) mandatory use of lime-treated aggregate in high elevation projects or problem pavement areas,
(2) increased asphalt content by reducing design voids from 4–7% to 3–5%,
(3) increased the IRS from 70% to 75%,
(4) reduced mix moisture from 1.0 in the road to 0.7% at plant,
(5) increased minimum air temperature requirements for compaction,
(6) required pneumatic rollers on all projects, and
(7) careful monitoring use of pick-up machines and storage silos.

TABLE 12 — *Changes in asphalt concrete specifications, 1974 through 1984 (Oregon State Highway Division).*

Parameters	1974	1984
Materials		
1. aggregates		
a. fracture	1 face, 60% by weight retained on ≥¼ in. screens	2 faces, 60% by weight retained on ≥No. 10 screens
b. soundness	≤18%	≤12%
c. stockpile	1 pile, 1974; 2 piles, 1977	1 coarse aggregate and 2 fine aggregate piles for plants without screens
2. recycled asphalt pavement (RAP)	no	≤20% in base course
3. mix moisture	≤3%	≤0.7% at plant
4. asphalt grades	AR	AR, AC, Pen
5. antistrip	chemical	hydrated lime and chemical
Mix Designs		
1. lab time	15 days	20 to 25 days
2. criteria		
a. % voids	4 to 7	3 to 5
b. stability	≥30	≥30
c. index retained strength	≥70%	≥75%
d. film thickness	visual	visual
3. field adjustment	none	2% CA, 1% FA, 0.5% P200
Construction		
1. air temperature	≥35°F	≥60°F, thin lifts ≥40°F, thick lifts
2. paving season	none	4/1 to 9/30; dense graded hot mix 5/1 to 8/31; open graded
3. compaction	92% relative maximum density	92% rice
4. windrow	none	complete pick up
5. construction joint	...	cut-back to uniform material
Equipment		
1. pneumatic tired rollers	optional	required
2. storage silos	...	batcher, rotating chute, or similar device
3. burners	...	aging index C ≥ 30
Acceptance and Payment		
1. system	...	quality assurance
a. process control	state	contractor
b. sampling	representative	random
c. acceptance	substantial compliance	statistical
2. payment		
a. penalty	minimal	$P = F(\overline{X} + aR - \underline{T}u)$ $P = F(T_L + aR - \overline{X}n)$
b. bonus	no	2 to 6%
c. escalation clause	no	yes-asphalt

NOTE: 1 in. = 2.54 cm and $t_c° = (t_F° - 32)/1.8$.

References

[1] Wilson, J. E. and Hicks, R. G., "Evaluation of Construction and Short-Term Performance Problems of Asphalt Pavement in Oregon," *Proceedings of the Association of Asphalt Paving Technologist,* Vol. 48, 1979, pp. 1–33.

[2] Hicks, R. G., Wilson, J. E., and Boyle, "Identification and Quantification of the Extent of Asphalt Stripping in Flexible Pavements in Oregon Phase/I," FHWA-OR-83-3, Federal Highway Administration, Washington, DC, March 1983.

[3] Santucci, L. E., Allen, D. D., and Coats, R. L., "The Effects of Moisture and Compaction on the Quality of Asphalt Pavements," prepared for presentation at the Annual Meeting of AAPT, University of Minnesota, Minneapolis, MN, Feb. 1985.

[4] Schlect, E. O., Jackson, N. C., and Walter, J. P., "Pavement Review for Oregon State Department," Report 188, Washington State Department of Transportation, Materials Office, Olympia, WA, Jan. 1984.

[5] "State Highway System Preservation Study," Oregon Department of Transportation, Jan. 1983.

[6] "Standard Method of Test for Effect of Water on Cohesion of Compacted Bituminous Mixtures," AASHTO T-165, Standard Specifications for Transportation Materials and Methods of Sampling and Testing, American Association of State Highway and Transportation Officials, 1982.

[7] Takallou, H., "Stripping of Asphalt Pavements: State of the Art," Masters Project, Transportation Engineering Report 84-6, Oregon State University, Corvallis, OR, 1984.

[8] Kim, O. K., Bell, C. A., and Hicks, R. G., "Effect of Moisture on Asphalt Pavement Life," prepared for presentation at the ASTM Symposium on Water Damage of Asphalt Pavements: Its Effect and Prevention, Williamsburg, VA, Dec. 1984.

[9] Lottman, R. P., "Predicting Moisture-Induced Damage to Asphaltic Concrete Field Evaluation Phase," NCHRP Report 246, National Cooperative Highway Research Program, Washington, DC, 1982.

[10] Witczak, M. W., "Design of Full Depth Asphalt Pavements," *Proceedings of the Third International Conference on Structural Design of Asphalt Pavements,* University of Michigan, Ann Arbor, MI, 1972.

Laboratory Evaluation of the Effects of Moisture, Antistripping, Additives, and Environmental Conditioning

Ok-Kee Kim,[1] Chris A. Bell,[1] and R. G. Hicks[1]

The Effect of Moisture on the Performance of Asphalt Mixtures

REFERENCE: Kim, O. K., Bell, C. A., and Hicks, R. G., **"The Effect of Moisture on the Performance of Asphalt Mixtures,"** *Water Damage of Asphalt Pavements: Its Effect and Prevention, ASTM STP 899,* B. E. Ruth, Ed., American Society for Testing and Materials, Philadelphia, 1985, pp. 51–72.

ABSTRACT: This paper presents the results of a laboratory study to investigate the effects of mixing moisture on mechanical properties of asphalt mixtures. The potential benefits of lime and Pavebond Special were also evaluated. The repeated load diametral test device was used to measure the mixture performance in terms of the resilient modulus, fatigue, and permanent deformation characteristics of laboratory specimens prepared with and without moisture (0, 1, and 3%) and with and without lime (1%) and Pavebond (0.5%). Mixtures were prepared that were representative of two projects for which considerable field data were available. One project utilized low quality and high absorptive aggregate and the other good quality aggregate. To evaluate the long-term durability of mixtures, they were tested before and after conditioning using the Lottman approach.

The test results showed that inferior performance occurred for mixtures with 3% moisture but was most pronounced in mixtures with higher air-void contents. However, the mixtures with low quality and high absorptive aggregate showed improved performance at 1% moisture content, associated with their lower air-void contents, which may be due to absorbed moisture preventing asphalt absorption and the higher asphalt content of these mixtures. The addition of lime resulted in distinct improvement of performance for moist specimens from the project, which had good quality aggregate, but high air-void contents. However, neither additive showed substantial benefit for moist samples from the project with low quality aggregate and low air-void content.

KEY WORDS: fatigue (materials), mixing, treatment, durability, compacting, mixing moisture, mixture performance, resilient modulus, permanent deformation

Introduction

Background

The Highway Division of the Oregon Department of Transportation and Oregon State University have collaborated for six years in evaluating the effects of

[1]Research assistant, assistant professor, and professor, respectively, Department of Civil Engineering, Oregon State University, Corvallis, OR 97331.

mix variations on asphalt pavement life [*1–3*]. To date, the results have led to improved specifications, but although mix moisture is now limited at or below 0.7% at the plant, this value is not based on rational research findings. Furthermore, no firm guidelines are available for selection of additives that may reduce the subsequent damage to mixtures containing moisture as a result of mixing with damp aggregates, a phenomenon primarily associated with the increased use of drum mixers [*2,3*].

The increased presence of residual moisture in asphaltic concrete mixtures caused by changes of materials and equipment is well known and has been the subject of considerable discussion recently. In the conference held during the Highway Research Board Annual Meeting in Jan. 1974 [*4*], there were many questions and considerable discussion about residual moisture. Among them were the following:

1. Are the moisture controls set at too low a level?
2. Is percent moisture a good measure of moisture effect?
3. Is the basic problem moisture or workability?
4. In the present situation, can we continue to say the drier the better?
5. If not, how high can we go and what should we control?

This study addresses some of the above questions and also considers the effect of two widely used additives. Mixtures were prepared that were representative of two projects about five years old in the State of Oregon for which considerable field data were available.

Purpose

The purpose of the study reported in this paper was to obtain a better understanding of the causes of the pavement problems associated with moisture, to develop relationships between pavement performance and the mixing moisture contents considering possible use of additives, and to develop mix design procedures that consider the moisture content and aggregate quality of asphalt mixtures. Such information could be useful in providing for limiting the moisture content of fresh mixtures and for selection of additives that may reduce the subsequent damage to mixtures containing moisture as a result of mixing with damp aggregates.

The specific objectives of this study were

(1) to evaluate the effect of moisture on mechanical properties of asphalt mixtures, such as resilient modulus, fatigue life, and permanent deformation,

(2) to evaluate the effect of additives for reducing the damage from moisture, and

(3) to provide guidelines to minimize the effect of moisture on pavement performance.

Research Approach

The research included tests on laboratory-prepared specimens representing two projects, North Oakland-Sutherlin and Warren-Scappoose. Following the standard Oregon Department of Transportation procedure [*5*], specimens 10.2 cm (4.0 in.) in diameter by 6.3 cm (2.5 in.) high were fabricated. Some modifications to this procedure were developed to incorporate moisture in some of the specimens. All specimens were tested in the diametral mode in accordance with ASTM Method for Indirect Tension Test for Resilient Modulus of Bituminous Mixtures (D 4123) for resilient modulus, fatigue life, and permanent deformation to evaluate the effects of different moisture contents (0, 1, and 3%) and additives (lime, Pavebond Special). Both as-compacted and conditioned asphalt mixtures were tested. Conditioned mixtures used the Lottman procedure [*6*], with subsequent testing providing a good indication of long-term durability.

Experiment Design

Projects Evaluated

North Oakland-Sutherlin:—This project is a section of Interstate 5 located approximately 19 km (12 miles) north of Roseburg, OR. Its overall length is 5.1 km (3.2 miles). The optimum asphalt content from the original mix design (1978) was 6.9% of an AR 8000 asphalt cement treated with 0.85% "Pavebond Special." The mix design was repeated for the laboratory tests, and because of differences in asphalt and aggregate, five years after construction, a different optimum asphalt content resulted as will be described below. The asphalt concrete base on this project was paved in Oct. through Dec. 1978, and showed problems of raveling and potholing shortly thereafter. An investigation performed by the Oregon Department of Transportation (ODOT) suggested that the reduced quality of the paving was basically the result of using varying amounts of low-quality aggregate in the mix [*2*]. The aggregate used in this project was a crushed submarine basalt containing seams of sulfate compounds of calcium, sodium, and magnesium. This aggregate showed high water absorption (up to 7% by weight). Soundness test results for produced aggregate used in the paving ranged from 4 to 39% loss for coarse aggregate and 11 to 48% loss for fine aggregate. Oregon currently (1984) limits the soundness test loss to 18%. Unfortunately, there were no data concerning the moisture content of cores. However, there was also some deviance from the specified aggregate gradation limits as shown in Fig. 1.

Warren-Scappoose:—This project is a section of the Columbia River Highway, located in Columbia county, OR. The overall length is 8 km (5 miles). The base course was constructed in 1979 and the wearing surface in 1980. The optimum asphalt content from the original mix design (1979) was 5.1% for the wearing surface and 5.7% for the base course, which used a good quality gravel aggregate. Again, the mix design was repeated for the laboratory tests, and

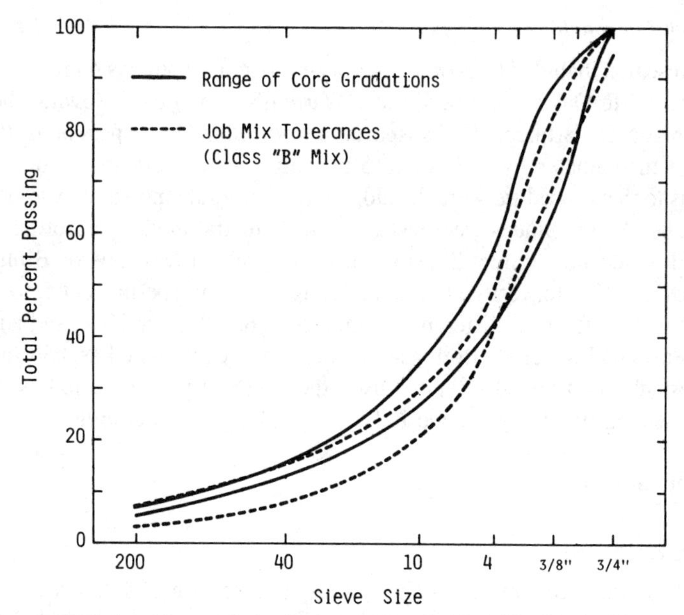

FIG. 1—*Core gradation for base layers: North Oakland-Sutherlin.*

because of differences in asphalt and aggregate, five years after construction, a different optimum asphalt content resulted as will be described below. The asphalt grade recommended was an AR 4000. Progressive pavement raveling and potholing were noticed in the base course during the months following construction [*3*]. The core data obtained for this project suggested that the reduction in pavement life resulted from high air-void content (10 to 15% for the wearing course and 6 to 11% for the base course) and variability in aggregate gradation as shown in Fig. 2. Also, the core data showed that the water content ranged from 0.64 to 1.0%, and poor adhesion or stripping was observed. Poor adhesion or stripping might have resulted from this water content.

Test Program and Methods

The variables considered in this study are presented in Table 1. All specimens were prepared to a target density of 96% of theoretical maximum, with the asphalt content and aggregate gradation established in the mix designs. The mix designs were repeated, since the aggregates and asphalts available, five years after the projects were constructed, were different to those used in the projects. The mix designs and specimen preparation for the laboratory study are outlined in subsequent sections.

A minimum of 12 specimens were prepared for each project for each one of the mix variables presented in Table 1 (a total of 168 specimens). Six specimens were tested as compacted, and six were tested after conditioning. All specimens were tested for resilient modulus, fatigue, and permanent deformation. All tests were

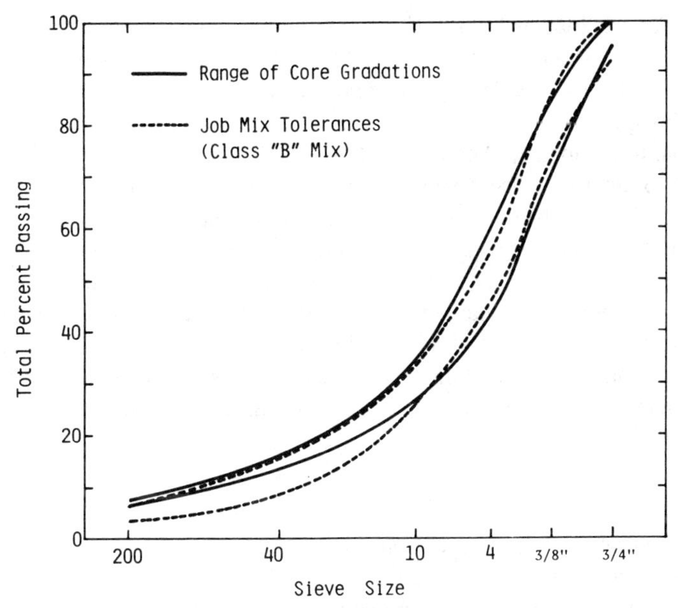

FIG. 2—*Core gradation for base layers: Warren-Scappoose.*

run at three different initial tensile strain levels ranging from 50 to 200 microstrain, resulting in two samples being used at each strain level.

The sample conditioning procedure was based on the moisture-induced damage test defined by Lottman [*6*]. In summary, (1) vacuum (66-cm [26-in.] mercury) saturate the specimens for 2 h, (2) place the saturated specimens in a freezer at −18°C (0°F) for 15 h, (3) place the frozen, saturated specimens in a warm water bath (60°C [140°F]) for 24 h, (4) place the specimens in a water bath at room temperature for 3 h, (5) dry the specimen at room temperature for 2 h, and (6) run the diametral test.

The resilient modulus, fatigue, and permanent deformation tests were performed using the repeated load diametral test apparatus. The test procedures employed are essentially the same as used in previous studies [*2,3*]. During the tests, the dynamic load duration was fixed at 0.1 s and the load frequency at

TABLE 1—*Range of mix variables considered in this study, X's.*

	Conditioning As-Compacted			Conditioned		
Antistrip Agent:	None	1% Lime	0.5% Pavebond Special	None	1% Lime	0.5% Pavebond Special
Moisture						
Standard (none)	X	X	X	X	X	X
1% moisture	X			X		
3% moisture	X	X	X	X	X	X

60 cpm. A static load of 4.5 kg (10 lb) was applied to hold the specimen in place. The tests were carried out at 22.5 ± 1.5°C (72.5 ± 2.7°F) and at 19.8 ± 1.5°C (67.6 ± 2.7°F) for the Warren-Scappoose and the North Oakland-Sutherlin projects, respectively.

The maximum load applied and the horizontal elastic tensile deformation were recorded to determine the resilient modulus. The resilient modulus M_R, initial horizontal elastic tensile strain ε_t, and vertical permanent strain ε_c were calculated from the equations suggested by Kennedy [7] using the specimen diameter of 10.2 cm (4.0 in.) and assumed Poisson's ratio of 0.35. Fatigue life is characterized by the number of load applications required to cause failure of the specimen. Attempts to relate the number of load applications to the specimen state of stress and strain showed that the best correlation exists between the tensile strain and the number of load applications, according to the following model [8,9]

$$N_f = K(1/\varepsilon_t)^m \tag{1}$$

where

N_f = number of load repetitions to failure,
K,m = regression constants, and
ε_t = initial horizontal elastic tensile strain.

The number of load repetitions to fatigue failure was defined as the number of repetitions required to cause a vertical crack approximately 0.65 cm (¼ in.) wide in the specimens. To stop the test at the specified level of specimen deformation, a thin aluminum strip was attached to the sides of the specimens, along a plane perpendicular to the plane formed by the load platens. When the specimen deformation exceeded a certain level, the aluminum strip broke and opened a relay, which shut off the test. Proper calibration of the length of the aluminum strip caused the test to stop for a specific specimen crack width of 0.65 cm (¼ in.).

For each test the accumulation of vertical permanent strain was also monitored during the fatigue life test using the controlled-load test method described by Monismith [10]. The relationship between vertical permanent strain and number of load repetitions may be expressed as follows

$$\varepsilon_c = I(N)^s \tag{2}$$

where

ε_c = compressive permanent vertical strain,
I,s = constants determined by regression analysis, and
N = number of load repetitions.

Mix Design

Standard mix designs for the laboratory study were carried out by Oregon State Highway Division Materials Section. The mix design method is a version of the Hveem approach (ASTM Test Methods for Resistance to Deformation and Co-

TABLE 2—*Mix design for laboratory prepared specimens: aggregate gradation, Class B.*

Sieve Size	Recommended Aggregate Gradation, %
1 in. (25.0 mm)	100
¾ in. (19.0 mm)	98
½ in. (12.5 mm)	87
⅜ in. (9.5 mm)	79
¼ in. (6.3 mm)	65
No. 10 (2.00 mm)	33
No. 40 (425 μm)	14
No. 200 (75 μm)	5.0

hesion of Bituminous Mixtures by Means of Hveem Apparatus [D 1560]), modified for Oregon conditions. Because of the availability of materials five years after construction, the mix designs for the laboratory study of moisture effects were different from the original mix designs for two projects described previously. As a result the asphalt contents used were 5.0 and 6.2% for the Warren-Scappoose and North Oakland-Sutherlin projects, respectively.

An Oregon Type B aggregate mix gradation (19-mm [¾-in.] nominal size) was used for this study (Table 2), the same as used in both projects [*2,3*]. The quality of the aggregates was similar to that used in the projects, viz, low soundness and high absorption characterized the North Oakland-Sutherlin project, and good quality gravel aggregate was used for the Warren-Scappoose project.

Specimen Preparation

The desired moisture contents for the asphalt concrete used in this study were from 0% moisture to a value of 3%, which has been demonstrated to be detrimental to asphalt concrete pavement. Hence, asphalt concrete specimens were prepared at 0, 1, and 3% moisture content, and with antistrip agents as indicated in Table 1. The asphalt concrete moist specimen preparation procedure was based on trial and error tests performed by the Oregon State Highway Materials Section. The following are the main steps:

1. Prepare specimen aggregate (1100 g).
2. Obtain the dry weight.
3. Soak aggregate in water overnight.
4. Heat the wet aggregate until the desired initial moisture content is obtained.
5. Add the asphalt cement required.
6. Mix and compact the specimen following ASTM Method for Preparation of Bituminous Mixture Test Specimens by Means of California Kneading Compactor (D 1561) procedure.

Results and Discussion

The results are presented and discussed below in two major parts: the effects of moisture on specimens with no additives and the effects of additives on

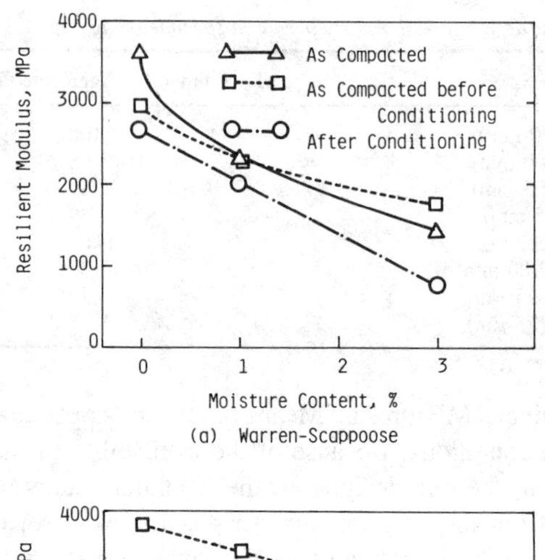

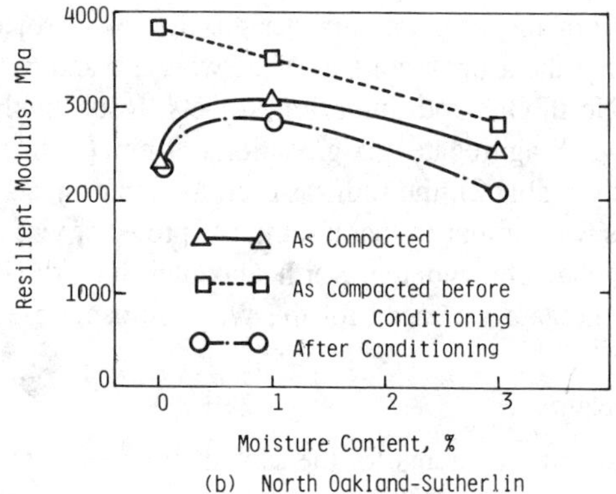

FIG. 3—*Effect of moisture on resilient modulus:* (a) *Warren-Scappoose and* (b) *North Oakland-Sutherlin.*

specimens with and without moisture. It should be noted that the moisture content refers to the amount incorporated during mixing, whereas "conditioning" refers to a test procedure [*6*] to evaluate durability of mixtures. In evaluating the results, the effect of moisture in "as-compacted" specimens (those involving no conditioning) is of much less significance than in "conditioned" specimens. The effect of conditioning is assessed by comparing properties measured with "as-compacted" and "conditioned specimens" such as a retained modulus value.

Effect of Moisture for Mixtures Without Additives

Modulus Results—Six specimens were tested as-compacted and six specimens were tested both before and after conditioning for each project for each mix variable set presented in Table 1. Because there was a difference of about seven weeks between tests with as-compacted specimens and the conditioning tests, two moduli values of as-compacted specimens per each mix variable set were obtained. The effect of moisture on modulus for both projects is shown in Fig. 3.

TABLE 3—*Specific gravity and air voids of laboratory specimens.*

Mix Moisture and/or Additive Amount	Maximum Specific Gravity	Specific Gravity	Air Voids, %	Water Intrusion[a], (%, by weight)
	WARREN-SCAPPOOSE			
0% moisture	2.487	2.284	8.16	...
1% lime	2.492	2.309	7.34	1.77
0.5% Pavebond	2.496	2.293	8.13	1.78
1% moisture	2.507	2.314	7.69	1.81
3% moisture	2.505	2.287	8.70	2.08
3% moisture/1% lime	2.493	2.315	7.14	1.19
3% moisture/0.5% Pavebond	2.508	2.256	10.05	1.82
	NORTH OAKLAND-SUTHERLIN			
0% moisture	2.493	2.285	8.34	1.60
1% lime	2.485	2.305	7.24	0.74
0.5% Pavebond	2.503	2.335	6.71	0.47
1% moisture	2.501	2.358	5.72	0.42
3% moisture	2.494	2.415	3.17	0.20
3% moisture/1% lime	2.478	2.388	3.63	0.47
3% moisture/0.5% Pavebond	2.486	2.430	2.25	0.17

[a]Water intrusion = $\frac{\text{weight after vacuum saturation} - \text{weight before vacuum saturation}}{\text{weight before vacuum saturation}} \times 100$

For the Warren-Scappoose project, modulus decreases with increasing moisture content. The average modulus of as-compacted specimens with 3% moisture content is 40% of the average modulus of as-compacted specimens with no moisture. Also, the retained modulus ratio (modulus after conditioning/modulus before conditioning) of specimens with 3% moisture content is only 0.43.

For the North Oakland-Sutherlin project, Fig. 3 shows that the average modulus of the as-compacted and conditioned specimens increases as the moisture content increases up to 1% and then decreases with increasing moisture. This behavior is due in part to the higher air voids (8.34%) for specimens without moisture compared to low air voids (5.72%) for specimens with 1% moisture, as shown in Table 3.

For the North Oakland-Sutherlin project, which used low-quality crushed rock aggregate, the average retained modulus ratio of specimens with no moisture (0.62) is lower than that for the Warren-Scappoose project (0.90), which used good quality gravel aggregate. Similar results were obtained in the previous study [*2,3*]. However, with 3% moisture content, the retained modulus ratio for the North Oakland-Sutherlin project (0.74) is about double that for the Warren-Scappoose project (0.42). This trend is attributed to different asphalt cement contents [*11*], resulting from the mix designs, (5.0% for the Warren-Scappoose project and 6.2% for the North Oakland-Sutherlin project), as well as the large difference of air voids shown in Table 3. The large difference in asphalt contents is due to the substantially different aggregates used in the projects, with the gravel aggregate in the Warren-Scappoose project requiring a much lower asphalt content.

The above results may be explained by considering the effects of asphalt quantity and aggregate quality on mixture resilient modulus. High stiffness (resilient modulus) is achieved by a high aggregate density, with pronounced interparticle friction and interlock. Aggregate density increases to a maximum (dependent on gradation) as the fluid content in a mixture increases until a maximum is reached, after which further addition of fluids (moisture and asphalt cement) causes a reduction in density. Such behavior is similar to that for soil and water combinations where maximum density occurs at an optimum water content. Maximum aggregate density may not correspond to maximum stiffness since asphalt contributes to mixture stiffness adversely once an aggregate density is reached that will mobilize interparticle friction and aggregate interlock. Addition of further asphalt will reduce aggregate interlock and interparticle friction. The presence of moisture at the time of mixing will contribute to additional fluids in the mixture and will therefore reduce the compactive effort necessary to achieve maximum aggregate density. For absorptive aggregates, free moisture in the aggregate leaves more asphalt available for coating the aggregate [*12*]. However, the bond between an asphalt cement and aggregate will usually be adversely affected by moisture, the severity depending on the aggregate and asphalt chemistry. This bond may not be very influential on mixture stiffness once the mixture becomes dry and stays dry [*11*]; however, the durability of the mixture may be significantly affected by wet-dry cycles or freeze-thaw cycles such as in the conditioning procedure used in this study [*6*].

The phenomena outlined above are exhibited in mixtures from both projects. For the mixtures from the Warren-Scappoose project, the modulus decreases with increased fluids content caused by loss of interparticle friction and interlock, since there was no increase in density of the mixture (Table 3), as would be expected with a gravel aggregate. The performance of this mixture becomes much worse as more moisture was present at mixing because of weak bonding, low asphalt content, and high voids in the mix.

For the mixtures from the North Oakland-Sutherlin project the tendency for increased modulus with a small addition of moisture is probably due to increased aggregate density affected by improved workability, as would be expected with crushed rock aggregate with greater potential for improved packing than a gravel aggregate. Additional moisture increases the aggregate density and mix density (Table 3), but reduces interparticle interlock and friction and therefore the resilient modulus. This mixture is more durable than that from the Warren-Scappoose project because of the increased density and higher asphalt content affected by improved workability.

An additional reason for the different behavior of the mixtures from the two projects is aggregate absorption. The marginal aggregate from the North Oakland-Sutherlin project was much more absorptive (up to 7% by weight), and this could be an advantage in mixtures using moist aggregates where water might prevent asphalt absorption and thus render more asphalt available for coating. It should be noted that following the conditioning procedure, mixtures from the

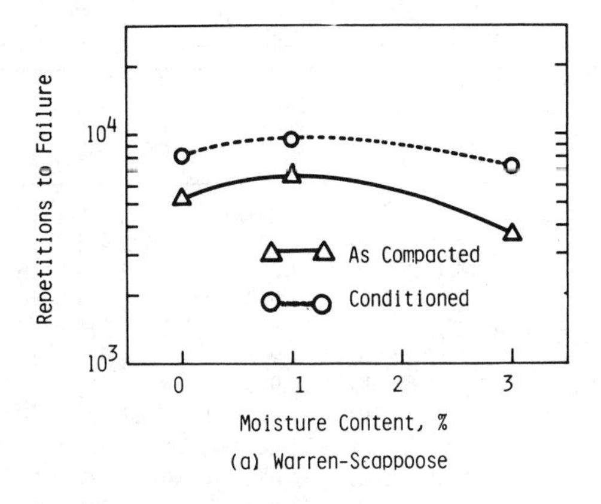

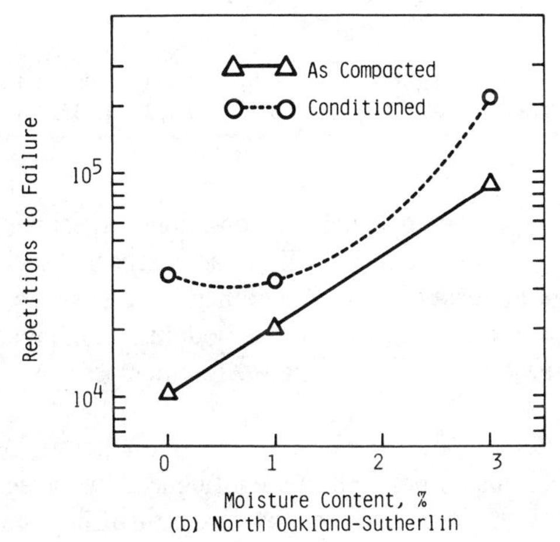

FIG. 4—*Fatigue life with moisture at* ε_t *of 100 microstrain:* (a) *Warren-Scappoose and* (b) *North Oakland-Sutherlin.*

Warren-Scappoose project had free water on the broken surface, whereas there was none on the surface of mixtures from the North Oakland-Sutherlin project, indicating higher aggregate absorption in the latter but lower mixture absorption caused by low air voids.

Fatigue Results—After the resilient modulus was measured, the fatigue life was determined at three different fixed initial tensile strain levels between 50 and 200 microstrains. The test was based on the controlled-load test method described in Ref *10*. The fatigue life of asphalt mixtures is a function of initial tensile strain and may be expressed by Eq 1, where the constants K and m shown in Table 3 were determined by linear regression analysis.

Figure 4 shows the number of load repetitions to failure at a measured strain of 100 microstrain for each project. These illustrate the results given in Table 4, and show quite clearly an "optimum" moisture content of 1% for the Warren-

TABLE 4—*Fatigue data* ($N_f = k[1/\varepsilon_t]^m$).

Mix Moisture and/or Additive Amount	As-compacted			Conditioned		
	K	m	r^2	K	m	r^2
			WARREN-SCAPPOOSE			
0% moisture	1.492×10^{-6}	2.387	0.995	5.636×10^{-4}	1.790	0.925
1% lime	3.101×10^{-9}	3.080	0.972	7.133×10^{-9}	3.010	0.978
0.5% Pavebond	1.268×10^{-8}	2.903	0.978	3.845×10^{-6}	2.303	0.959
1% moisture	2.729×10^{-6}	2.344	0.912	1.270×10^{-7}	2.718	0.929
3% moisture	6.821×10^{-5}	1.932	0.953	4.908×10^{-9}	3.044	0.919
3% moisture/1% lime	4.989×10^{-8}	2.921	0.968	5.412×10^{-6}	2.321	0.911
3% moisture/0.5% Pavebond	9.186×10^{-5}	1.922	0.975	1.004×10^{-2}	1.474	0.941
			NORTH OAKLAND-SUTHERLIN			
0% moisture	2.451×10^{-7}	2.658	0.999	1.637×10^{-16}	5.083	0.976
1% lime	6.541×10^{-9}	3.109	0.941	1.295×10^{-8}	3.023	0.905
0.5% Pavebond	2.498×10^{-11}	3.724	0.968	4.797×10^{-20}	6.055	0.960
1% moisture	6.717×10^{-11}	3.621	0.944	4.617×10^{-14}	4.464	0.826
3% moisture	3.604×10^{-18}	5.599	0.861	3.102×10^{-14}	4.712	0.927
3% moisture/1% lime	1.716×10^{-18}	5.675	0.944	2.982×10^{-4}	2.075	0.880
3% moisture/0.5% Pavebond	4.210×10^{-17}	5.307	0.973	1.735×10^{-14}	4.797	0.917

Scappoose project, for as-compacted and conditioned specimens. For the North Oakland-Sutherlin project, the fatigue life of as-compacted specimens decreases as moisture content increases up to 1% and then increases with increasing moisture. This fatigue result is the inverse of the modulus result presented in Fig. 3*b*. After conditioning, however, the fatigue life increases as moisture content increases.

The above results can be explained in terms of the stiffness and air-void contents of the materials, which in turn are influenced by the aggregate type and asphalt content. For the Warren-Scappoose project stiffness decreased with increasing moisture content (Fig. 3), and the air-void content was lowest at 1% moisture (Table 3). The substantial loss in modulus should lead to longer fatigue lives at higher moisture content on the basis of the controlled load test, but the air voids effect plus the probable weaker bond between asphalt and aggregate (caused by the free water between them) lead to the shorter fatigue lives at higher moisture content for this project.

For the North Oakland-Sutherlin project, the much longer fatigue lives at higher moisture content are a result of lower air-void contents. For both projects, fatigue lives were longer after conditioning for all moisture conditions (Fig. 4) because of the reduced stiffness that occurred in all cases (Fig. 3).

Deformation Results—The effect of moisture on permanent deformation of mixtures at initial tensile strain of 100 microstrain for the Warren-Scappoose project and 150 microstrain for the North Oakland-Sutherlin project is presented in Figs. 5 and 6, respectively. For the Warren-Scappoose project, the as-compacted specimens with 3% moisture again performed the worst (less repetitions are required to reach a fixed compressive strain), and specimens with 0%

(a) As Compacted

(b) Conditioned

FIG. 5—*Effect of moisture on permanent deformation of Warren-Scappoose project at* ε_t *of 100 microstrain:* (a) *as-compacted and* (b) *conditioned.*

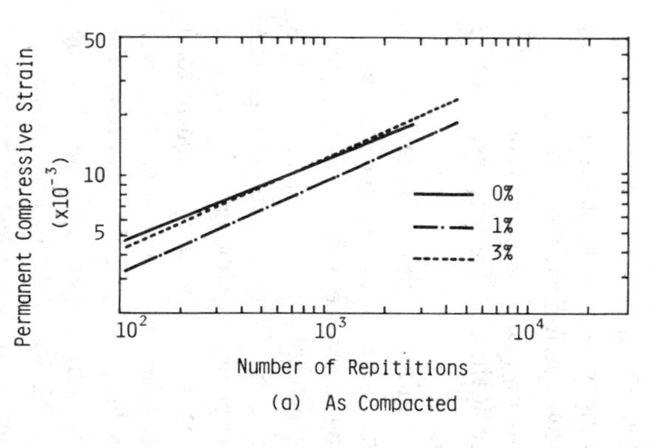

(a) As Compacted

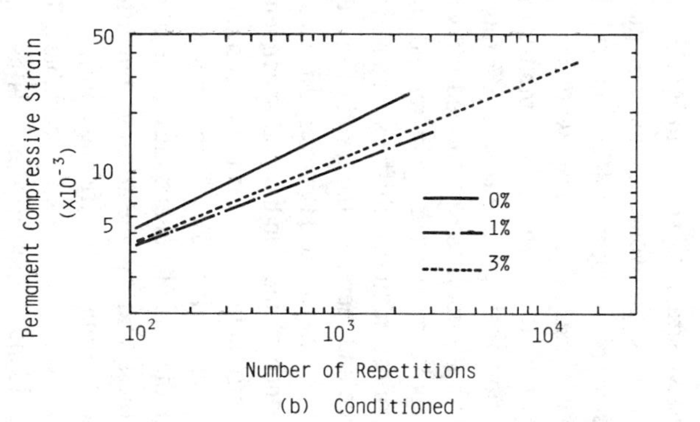

(b) Conditioned

FIG. 6—*Effect of moisture on permanent deformation of North Oakland-Sutherlin project at* ε_t *of 150 microstrain:* (a) *as-compacted and* (b) *conditioned.*

moisture performed the best, (more repetitions are required to reach a fixed compressive strain) as shown in Fig. 5. After conditioning, specimens with 1% moisture gave the worst results. For the North Oakland-Sutherlin project, as-compacted specimens with 1% moisture (showing the highest modulus in Fig. 3*b*) performed best, while those with 3% moisture performed worst in the high compressive strain range, as demonstrated in Fig. 6. After conditioning, the results show that specimens with 1% moisture again gave the best results while those with 0% moisture performed the worst.

The above results are again related to the stiffness, air voids, and asphalt content of the mixtures. Stiffness is a major influence on permanent deformation resistance and therefore the trend of decreasing resistance with decreasing stiffness is to be expected. However, this was not the case for all of the results. The performance of the conditioned specimens from the Warren-Scappoose project was slightly better than for the as-compacted specimens although they were of lower stiffness (Fig. 3). This anomaly may be due to the comparison method based on the strain level and the diametral mode of testing, which is most suitable for stiffness and fatigue testing, but after a large number of repetitions the permanent strain is significantly influenced by various geometric and loading factors as well as the mixture variables. The behavior of the mixtures from the North Oakland-Sutherlin project is as expected with the stiffest mixtures performing best.

Effect of Additives

Modulus Results — A major part of the testing program was aimed at evaluating the effect of lime (1%) and Pavebond Special (0.5%) on the performance of asphalt mixtures. The effect of additives on modulus for mixtures with no moisture is presented in Fig. 7. For both projects, introduction of additives increases the modulus of as-compacted specimens very slightly. After conditioning, specimens with additives obtained a much higher modulus than the "no treatment" specimens and specimens with 1% lime obtained the highest modulus.

With 3% moisture, the increase of modulus and retained modulus ratio by lime is substantial for the Warren-Scappoose project, and there is an improvement in retained modulus as shown in Fig. 8. For the North Oakland-Sutherlin project the moduli of as-compacted and conditioned specimens with no additives are about the same as those with lime or Pavebond Special.

For the Warren-Scappoose project, without moisture there is little benefit from use of additives except with conditioned specimens. However, for specimens with moisture, the effect of lime is superior to that of Pavebond Special. For the North Oakland-Sutherlin project, there is little benefit shown from use of additives. Ishai and Craus [*13*] emphasized that the contribution of the hydrated lime to adhesion is mobilized only in the presence of water. The above results confirm their conclusion since the best improvement of performance was obtained in the Warren-Scappoose specimens with 3% moisture. There was little benefit in the North Oakland-Sutherlin project because of the superior durability of the mixture.

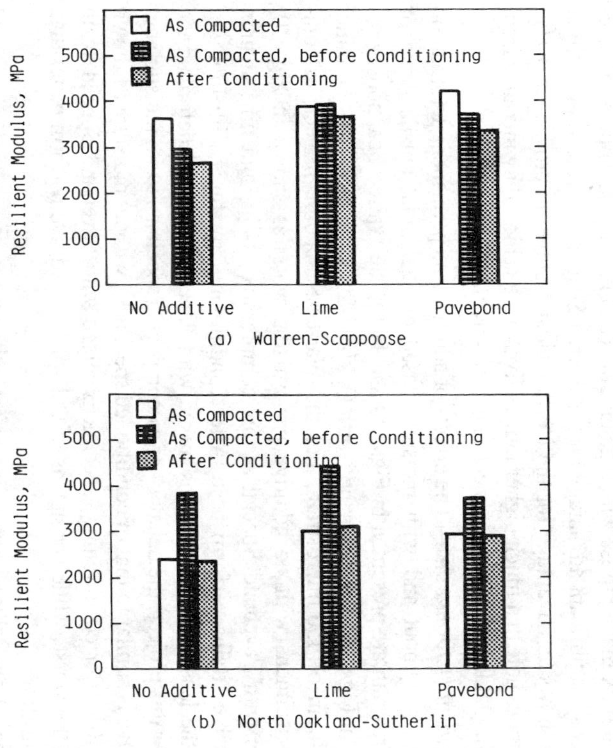

FIG. 7—*Effect of additives without moisture on resilient modulus: (a) Warren-Scappoose and (b) North Oakland-Sutherlin.*

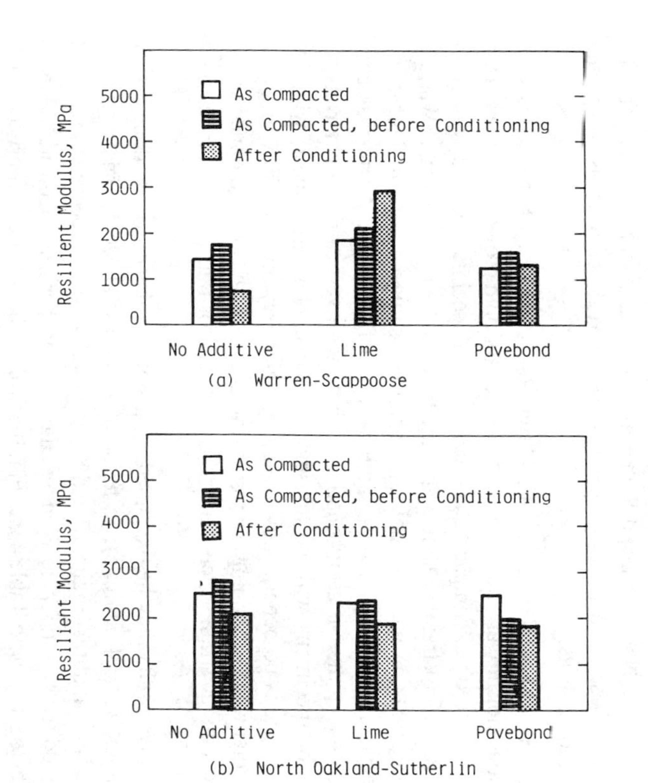

FIG. 8—*Effect of additives with moisture (3%) on resilient modulus: (a) Warren-Scappoose and (b) North Oakland-Sutherlin.*

It can be seen that there is a partial stiffening effect on the specimens without moisture for the North Oakland-Sutherlin project and with moisture for the Warren-Scappoose project during a period of seven weeks between as-compacted and conditioned tests.

Fatigue Results—The effects of additives on fatigue life of specimens with and without moisture are summarized in Table 4 for both projects. A comparison between fatigue of as-compacted and conditioned specimens mixed with additives at 100 microstrain is shown in Fig. 9. It is clear that for the Warren-Scappoose mixtures with no moisture, additives show no benefit, but, when moisture is present, lime shows significant benefit. For the North Oakland-Sutherlin project, there is clear benefit from both additives, particularly Pavebond Special, for specimens with no moisture. However, for this project, with 3% moisture, there is no benefit shown with lime and very little benefit with Pavebond Special when examining the results for as-compacted or conditioned specimens. Results at strain levels other than 100 microstrain are similar to those shown in Fig. 9.

The above results may be explained, in a similar manner to that done earlier for mixtures without additives, in terms of modulus, air voids, and degree of coating of the mixtures. Figures 7 and 8 show moduli for the specimens with and without additives, and Table 3 shows the air-void contents. Consideration of these data together with Table 4 and Fig. 9 confirms the previous observation that the additives are only likely to be of significant benefit in mixtures with questionable durability, that is, those with high air voids and low asphalt content. Hence, lime shows a significant advantage for the Warren-Scappoose project, but neither additive shows an advantage for the North Oakland-Sutherlin project. It is also significant to note that much higher fatigue lives were achieved by mixtures from the North Oakland-Sutherlin project under all circumstances.

Permanent Deformation—The effects of additives on permanent deformation of mixtures without and with moisture based on the initial tensile strain of 100 microstrain are presented in Figs. 10 and 11 for the Warren-Scappoose project and in Figs. 12 and 13 for the North Oakland-Sutherlin project at the initial tensile strain of 150 microstrain. For the Warren-Scappoose project, as-compacted specimens with no additive perform the worst as shown in Fig. 10. Mixtures with Pavebond Special at low compressive strains and lime at high compressive strains perform best. After conditioning, mixtures with lime deformed the least, while specimens mixed with Pavebond Special deformed the most. For this project, the behavior of as-compacted specimens with moisture but no additive is similar to that of conditioned specimens without moisture as shown in Fig. 10. After conditioning, specimens mixed with Pavebond Special again perform the worst, while specimens mixed with no additive perform better than those with lime.

For the North Oakland-Sutherlin project, without moisture, specimens mixed with lime show the best performance for both as-compacted and conditioned specimens as shown in Fig. 12. With moisture, specimens mixed with lime again

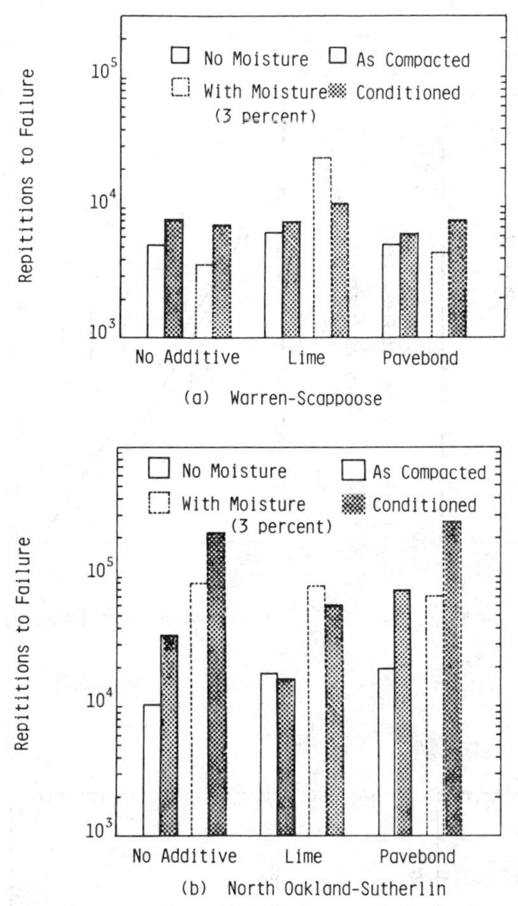

FIG. 9—*Fatigue life with additives at* ε_t *of 100 microstrain:* (a) *Warren-Scappoose and* (b) *North Oakland-Sutherlin.*

show the best performance for both as-compacted and conditioned specimens as shown in Fig. 13. With and without moisture specimens mixed with no additives show the worst performance. As mentioned earlier, deformation results at high strain levels should be treated with caution when obtained with a diametral device. Indeed, as in the evaluation of results for mixtures without additives, the deformation results do not compare well with other data.

General Discussion

The above discussion indicates that fatigue and deformation results may be explained in terms of the stiffness results, with a reduction in stiffness tending to cause prolonged fatigue life (because of the comparison based on strain level) but a decrease in deformation resistance. It is therefore essential to recognize that the fatigue results should not be considered by themselves as an indicator of the effect

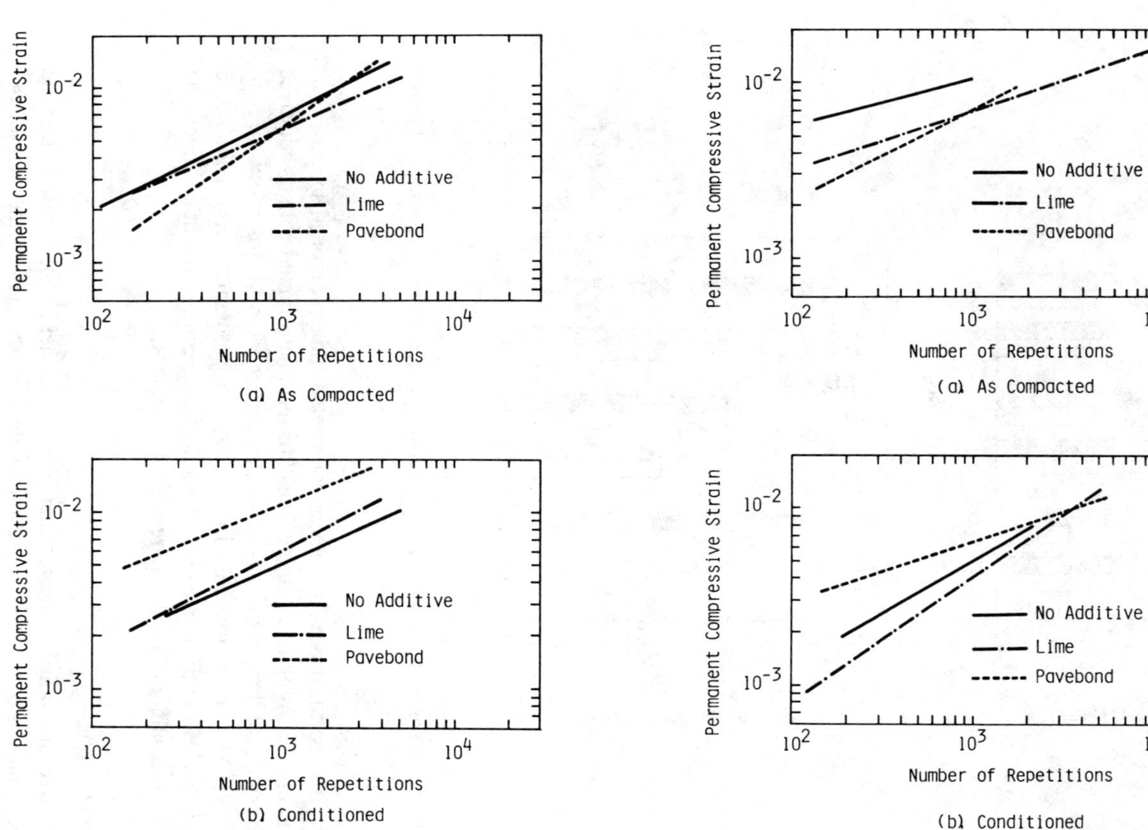

FIG. 10—*Effect of additives with moisture (3%) on permanent deformation of Warren-Scappoose project at* ε_t *of 100 microstrain:* (a) *as-compacted and* (b) *conditioned.*

FIG. 11—*Effect of additives without moisture on permanent deformation of Warren-Scappoose project at* ε_t *100 microstrain:* (a) *as-compacted and* (b) *conditioned.*

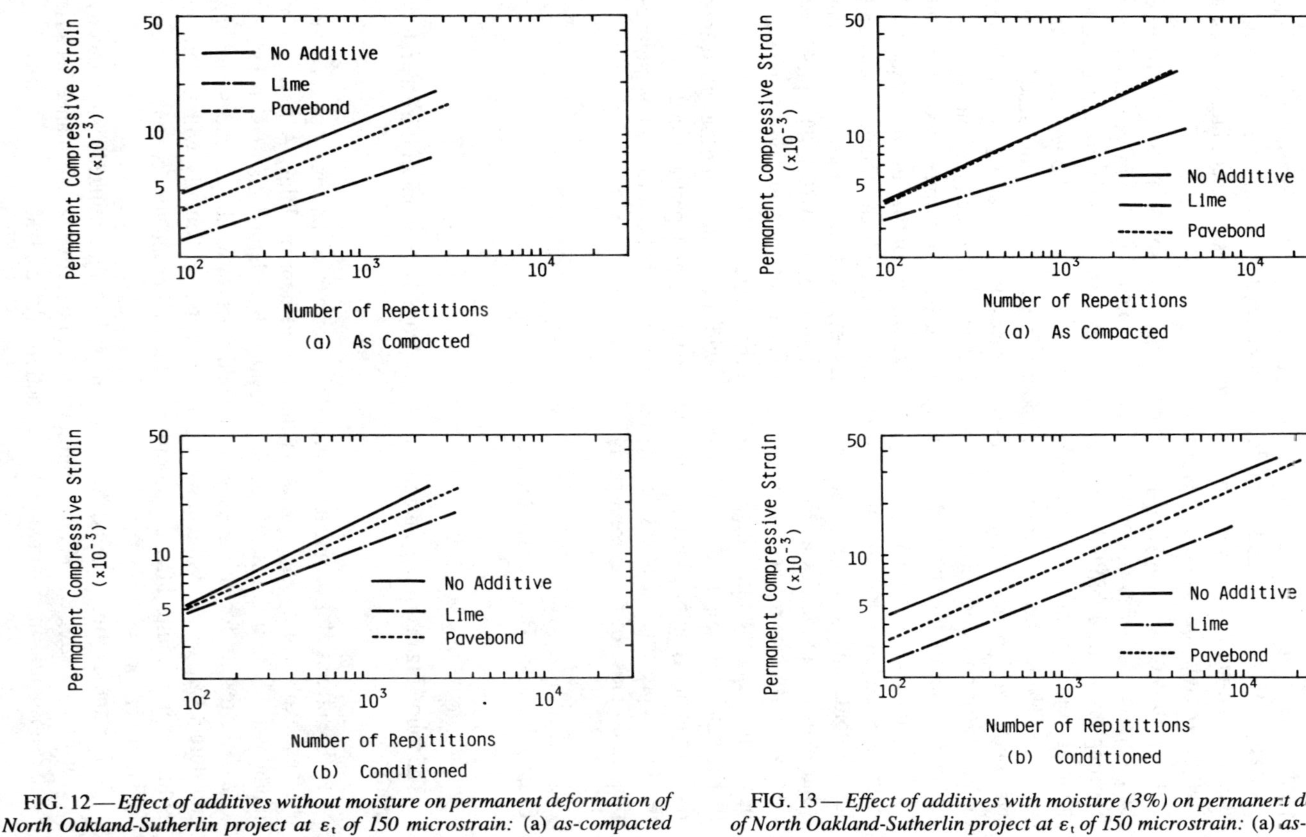

FIG. 12—*Effect of additives without moisture on permanent deformation of North Oakland-Sutherlin project at ε_t of 150 microstrain:* (a) *as-compacted and* (b) *conditioned.*

FIG. 13—*Effect of additives with moisture (3%) on permanent deformation of North Oakland-Sutherlin project at ε_t of 150 microstrain:* (a) *as-compacted and* (b) *conditioned.*

of moisture and additives on pavement life, since they do not reflect differences in mixture stiffnesses. Rather a study such as that reported by Bell et al [14] should be done to determine the effect of changes in stiffness on the distress parameters associated with pavement failure. In particular, determination of the critical tensile strain in the asphalt layer can be used in conjunction with fatigue data and the number of load repetitions to failure assessed accordingly. Such a study is outside the scope of this paper.

The results of this study partially confirm the assessment of the performance of the projects that was observed after construction [2,3]. In particular, the raveling problems exhibited in the Warren-Scappoose project were attributed to high air void content following construction (10 to 15% for the wearing course and 6 to 11% for the base course) and poor adhesion or stripping caused by the water content (0.64 to 1.0%). This study confirms that high air voids contributed to poorly performing mixtures particularly if mix moisture was initially high.

For the North-Oakland Sutherlin project, the laboratory mixtures performed better than expected compared to the in-service performance. This was associated with the excellent durability of the mixtures caused by high asphalt content and low air voids, and good mechanical properties, even with mix moisture exceeding what is now regarded as a tolerable amount in Oregon (0.7%). The problems observed in service may not have been associated with the low-quality aggregate used, but rather the fairly high air voids in the mixture [2].

The results of this study are in keeping with the philosophy that good mix design and high compaction (low air voids) result in a high performance mix. The results of a previous study at Oregon State University [1] confirmed this philosophy, as do those of the current study.

Conclusions and Recommendations

Conclusions

The following major conclusions are drawn from the findings of this laboratory study:

1. The effects of mix moisture and additives on asphalt pavement performance are best explained by the effects on mixture resilient modulus. A lowering of modulus tends to increase fatigue life, but decrease deformation resistance.

2. The results for mixtures with no additives showed that excess mixing moisture (3%) was detrimental for both projects, particularly with regard to reduction in modulus after conditioning in the higher air-void content mixtures (Warren-Scappoose project).

3. For the North Oakland to Sutherlin project that utilized a low-quality crushed rock aggregate, 1% moisture resulted in improved performance because of the substantial improvement in compaction that resulted.

4. For mixtures with additives, the detrimental effect of 3% moisture on the resilient modulus was substantially reduced by use of lime (1%) in the Warren-Scappoose project, with Pavebond Special (0.5%) showing limited benefit. Neither additive showed substantial benefit in the North Oakland-Sutherlin project.

5. Additives are of limited benefit in mixtures with high density and which achieve good performance without additives, such as the North Oakland-Sutherlin project. Although this project used a low-quality aggregate, the mix design produced a high performance mixture. Additives, particularly lime, were of substantial benefit in the Warren-Scappoose project, which had a high air-void content and low retained modulus without the additives.

Recommendations

The following recommendations can be made from the results of this phase of the study:

1. Since the limit of acceptable moisture content could vary from one project to another (depending on the absorption of the aggregate), limiting values could also be established after a regular mix design by additional tests at the optimum asphalt content with moisture contents varying up to the maximum absorption of the aggregate.

2. It is recommended to use as much asphalt cement as possible to maximize durability and minimize the damage from moisture, while ensuring adequate resistance to permanent deformation.

3. Mix density should be maximized, commensurate with good mix design practice, to achieve good mechanical properties and durability.

4. Additives should be selected after appropriate tests have established values such as the retained modulus ratio.

5. It is necessary to establish test methods to determine the effect of residual moisture on mixtures. The procedures employed in this study together with the specimen preparation techniques are a good starting point.

6. Since the effect of moisture and effectiveness of additives is a complex function of asphalt, aggregate, and additive interactions, it is necessary to establish a simple test to determine the compatibility of these components.

Acknowledgments

This report presents the results from the first phase of a two-phase Highway Planning and Research (HP and R) study conducted by Oregon State Highway Division and Oregon State University in cooperation with the Federal Highway Administration.

The contribution of James E. Wilson, Glen Boyle, and their staff in obtaining materials and preparing mix designs was invaluable. The authors are indebted to Laurie Campbell of the Engineering Experiment Station, Oregon State University, who typed the manuscript.

Disclaimer

The contents of this paper reflect the views of the authors who are responsible for the facts and accuracy of the data presented. The contents do not necessarily reflect the official views or policies of either the Oregon State Highway Division or Federal Highway Administration.

References

[1] Puangchit, P., Hicks, R. G., Wilson, J. E., and Bell, C. A., "Development of Rational Pay Adjustment Factors for Asphalt Concrete," *Transportation Research Record 911*, Transportation Research Board, Washington, DC, 1983, pp. 70–79.

[2] Walter, J. L., Hicks, R. G., and Wilson, J. E., "Impact of Variation in Material Properties on Asphalt Pavement Life—Evaluation of North Oakland-Sutherlin Project," FHWA-OR-81-4, Interim report to Federal Highway Administration and Oregon Department of Transportation, Salem, OR, Dec. 1981.

[3] Walter, J. L., Hicks, R. G., and Wilson, J. E., "Impact of Variation in Material Properties on Asphalt Pavement Life—Evaluation of Warren-Scappoose Project," FHWA-OR-81-7, Interim report to Federal Highway Administration and Oregon Department of Transportation, Salem, OR, Dec. 1981.

[4] Transportation Research Board, "Second Look at Moisture Restrictions in Hot-Mix Plant Operations and Construction," *Transportation Research Circular 262*, Transportation Research Board, Washington, DC, 1983.

[5] "Laboratory Manual of Test Procedures," Laboratory Manual Vol. 1, Oregon Department of Transportation, Highway Division, Material and Research Section, Salem, OR, March 1978.

[6] Lottman, R. P., "Predicting Moisture-Induced Damage to Asphaltic Concrete," U. I. Project 677-K297, University of Idaho, Moscow, ID, Feb. 1979.

[7] Kennedy, T. W., "Characterization of Asphalt Pavement Materials Using the Indirect Tensile Test," *Proceedings*, Association of Asphalt Paving Technologists, Vol. 46, 1977, pp. 132–150.

[8] Pell, P. S., "Characterization of Fatigue Behavior," Special Report 140, Highway Research Board, Washington, DC, 1973, pp. 49–65.

[9] Witczak, M. W., "Fatigue Subsystem Solution for Asphalt Concrete Airfield Pavements," Special Report 140, Highway Research Board, Washington, DC, 1973, pp. 112–130.

[10] Monismith, C. L., "Influence of Shape, Size, and Surface Texture on the Stiffness and Fatigue Response of Asphalt Mixtures," Special Report 109, Highway Research Board, Washington, DC, 1970, pp. 4–11.

[11] Schmidt, R. J. and Graf, P. E., "The Effect of Water on the Resilient Modulus of Asphalt-Treated Mixes," *Proceedings*, Association of Asphalt Paving Technologists, Vol. 41, 1972, pp. 118–162.

[12] Schmidt, R. J., Kari, W. J., Bower, H. C., and Hein, T. C., "Behavior of Hot Asphaltic Concrete Under Steel-Wheel Rollers," *Highway Research Board Bulletin 251*, Washington, DC, 1960, pp. 18–37.

[13] Ishai, I. and Craus, J., "Effect of Filler on Aggregate-Bitumen Adhesion Properties in Bitumen Mixtures," *Proceedings*, Association of Asphalt Paving Technologists, Vol. 46, 1977, pp. 228–258.

[14] Bell, C. A., Hicks, R. G., and Wilson, J. E., "Effect of Percent Compaction on Asphalt Pavement Life," *Placement and Compaction of Asphalt Mixtures, STP 829*, F. T. Wagner, Ed., American Society for Testing and Materials, Philadelphia, 1984, pp. 107–130.

Dennis W. Gilmore, [1] *James B. Darland, Jr.* [1] *Larry M. Girdler,* [1] *Lewell W. Wilson,* [1] *and James A. Scherocman* [1]

Changes in Asphalt Concrete Durability Resulting from Exposure to Multiple Cycles of Freezing and Thawing

REFERENCE: Gilmore, D. W., Darland, J. B., Jr., Girdler, L. M., Wilson, L. W., and Scherocman, J. A., **"Changes in Asphalt Concrete Durability Resulting from Exposure to Multiple Cycles of Freezing and Thawing,"** *Evaluation and Prevention of Water Damage of Asphalt Pavement Materials, ASTM STP 899,* B. E. Ruth, Ed., American Society for Testing and Materials, Philadelphia, 1985, pp. 73–88.

ABSTRACT: Concern over water-induced damage in asphalt concrete pavements has resulted in the development of several new moisture susceptibility detection tests. This report compares two modifications of the water conditioning procedure for asphalt concrete specimens as described in the National Cooperative Highway Research Program (NCHRP) Reports 192 and 246. These modifications incorporate longer water exposure conditions for investigating the long-term durability of antistripping additives and paving mixtures.

Significant and progressive changes in tensile strength were found to occur when asphalt concrete specimens were repeatedly exposed to freeze-thaw cycles. Factors, such as aggregate source, additive type, and additive quantity, influence the rate of strength loss during cyclical conditioning, and these rates are valuable in material selection. In particular, the sensitivity of cyclical conditioning to additive effectiveness was utilized to develop and select improved additives.

KEY WORDS: bituminous cements, bituminous concretes, additives, diazonium salts, tensile strength, X-ray fluorescence, antistripping, continuous soaking, fatigue life, freeze-thaw cycle, moisture susceptibility, resilient modulus, tensile strength ratio (TSR), TSR loss rate, vacuum saturation

Extensive research over the past decade produced several new methods for measuring moisture sensitivity of asphalt concrete mixtures and the effectiveness of additive treatments in these mixtures [*1*]. One of the more widely utilized of these methods is the indirect tension test as developed by Professor Robert Lottman [*2*,*3*]. Careful study by Lottman determined that a moisture conditioning

[1]Research manager, research technician, research chemist, senior research technician, and marketing director, respectively, MORTON THIOKOL, Inc., Carstab Division, 4800 West St., Cincinnati, OH 45215.

process of vacuum saturation followed by a freeze-thaw cycle (FTC) leads to approximately the same amount and the same type of moisture damage as does five years of pavement service. Other researchers also correlated lab results from the Lottman method with field observations and report the lab procedure to be equivalent to 10 to 15 natural freeze-thaw cycles (FTC) [*4*]. On the other hand, some investigators feel the Lottman conditioning process is too severe and have proposed procedural variations that eliminate the freezing step and reduce testing time [*5*].

From the standpoint of asphalt additive durability and endurance, we feel that 1 FTC may not represent the full potential for long-term moisture damage. Consequently, Lottman's procedure was modified to create a more severe water conditioning environment in order to permit better study of the long-term effectiveness of additives. The procedures described herein may not be practical for routine laboratory use. Significant potential is, however, demonstrated for using these procedures in the research and development of improved additives.

Procedure

Generation of moisture damage in asphalt concrete specimens that is greater than 1 FTC was initially explored by two different methods. In Method A the specimens (6 by 10.2 cm [2.5 by 4.0 in.]) were thawed at 140°F (60°C) for time periods that extended well beyond the normal 24 h. For Method B the entire freeze-thaw cycle process was repeated several times. To permit convenient scheduling after the first FTC, the freeze time in each cycle was extended from 15 to 24 h. This variation produced no noticeable change in results when compared to the normal 15-h freeze time. The thaw time of 24 h/cycle remained constant throughout Method B. The progress of moisture damage in each conditioning method was measured by changes in tensile strength.

Several dozen different asphalt concrete mixtures were examined during the course of this study. Most of these mixtures were prepared and compacted (specimens are fabricated according to ASTM Test Method for Resistance to Plastic Flow of Bituminous Mixtures Using Marshall Apparatus [D 1559]) in the laboratory in order to determine the affect of several variables on mixture sensitivity to prolonged moisture exposure. Some field plant-mix samples were also obtained, and results with these materials are described.

Details concerning the mechanical tests for tensile strength, resilient modulus (ASTM Method for Indirect Tension Test for Resilient Modulus of Bituminous Mixtures [D 4123]), and fatigue life are adequately described elsewhere [*2,3,6*]. The quantitative detection of hydrated lime in asphalt concrete mixtures was accomplished with X-ray fluorescence spectroscopy (XRF). Appropriate calibration curves for lime content were developed by varying the known amount of lime present in lab-prepared asphalt concrete mixtures.

The presence of polyamine additive in the asphalt concrete was determined by extractive isolation with aqueous hydrochloric acid and hydrocarbon solvents followed by reaction with 2,4-dinitrobenzenediazonium tetrafluoroborate. The polyamine-diazonium salt reaction product is characterized by ultraviolet (UV) visible chromophore (γ maximum = 490 nm), and the intensity of this chromophore is a quantitative measure of free polyamine in asphalt concrete. Appropriate calibration samples were generated by varying the known amount of polyamine in the mixtures.

Results and Discussion

Method Selection

Initial investigations of Methods A and B were conducted with two dense-graded asphalt concrete mixtures and two commercial liquid antistripping additives. The limestone/sand aggregate (from Austin, TX) is reported to be slightly moisture sensitive and the granite gneiss (Lithonia, GA) is considered to be very susceptible to moisture damage. Specific composition and mechanical property data are presented in Table 1.

The tensile strength ratio (TSR) values for Mixtures 1 through 6 after vacuum saturation only, one FTC and FTC plus extended soaking at 60°C (140°F) are presented in Fig. 1. Stripping was visually detected in Mixtures 1 and 5 at all stages of moisture conditioning. This stripping was especially pronounced for Mix 1 after one FTC.

The commercial asphalt additives increased the TSR values and reduced the stripping for both aggregate systems regardless of the extent of moisture conditioning. The siliceous aggregate is slightly sensitive to additive type but is more responsive to changes in additive use level. Despite better performance for Mix 5 compared to Mix 1, the limestone/sand aggregate is not benefited by the amidoamine additive as substantially as is the siliceous aggregate. This difference reflects the structure of the amidoamine additive, which is designed to be more responsive to siliceous (acidic) rather than calcareous (basic) aggregates [*5*,*7*].

Further examination of Fig. 1 indicates that most of the water damage for a given mixture (TSR loss) normally occurs within the freeze-thaw cycle portion of Method A. The relative drop in TSR after the FTC varies from mixture to mixture, and in some cases (for example, Mixtures 2, 3, 4, and 6) this drop is inconsequential even after two weeks of soaking. Mixture 5 suffers the most damage during extended soaking, but this soaking-only damage does not exceed the TSR loss caused by 1 FTC. Based on the results presented in Fig. 1, one would expect good long-term durability for asphalt concrete Mixtures 2, 3, 4, and 6 if 60°C (140°F) soaking represents the most severe form of water conditioning for in-service pavements.

Tensile strength changes for Mixtures 1 through 6 resulting from repetitive FTC conditioning (Method B) are presented in Fig. 2. Unlike for Method A, specimens conditioned by Method B experience considerable moisture damage

TABLE 1—*Data for granite and limestone/sand asphalt concrete mixtures.*

Mix Number	Aggregate[a]	Additive	Percent Air Voids[b]	Dry Tensile Strength, psi	Percent Saturation[c]
1	granite	none	6.3	107	84
2	granite	0.25% polyamine	6.1	106	80
3	granite	0.50% polyamine	6.0	103	76
4	granite	0.50% amidoamine	6.0	101	81
5	limestone/sand	none	7.0	71	84
6	limestone/sand	0.50% amidoamine	7.2	68	88

[a]All mixes were prepared with an AC-20 asphalt cement at 5.5 weight % for the granite and 5.1 weight % for the limestone/sand.

[b]ASTM Test Method for Theoretical Maximum Specific Gravity of Bituminous Paving Mixtures (D 2041), Test Method for Bulk Specific Gravity of Compacted Bituminous Mixtures Using Saturated Surface-Dry Specimens (D 2726), and Test Method for Percent Air Voids in Compacted Dense and Open Bituminous Paving Mixtures (D 3203).

[c]Saturation was conducted at a vacuum of 100-mm mercury.

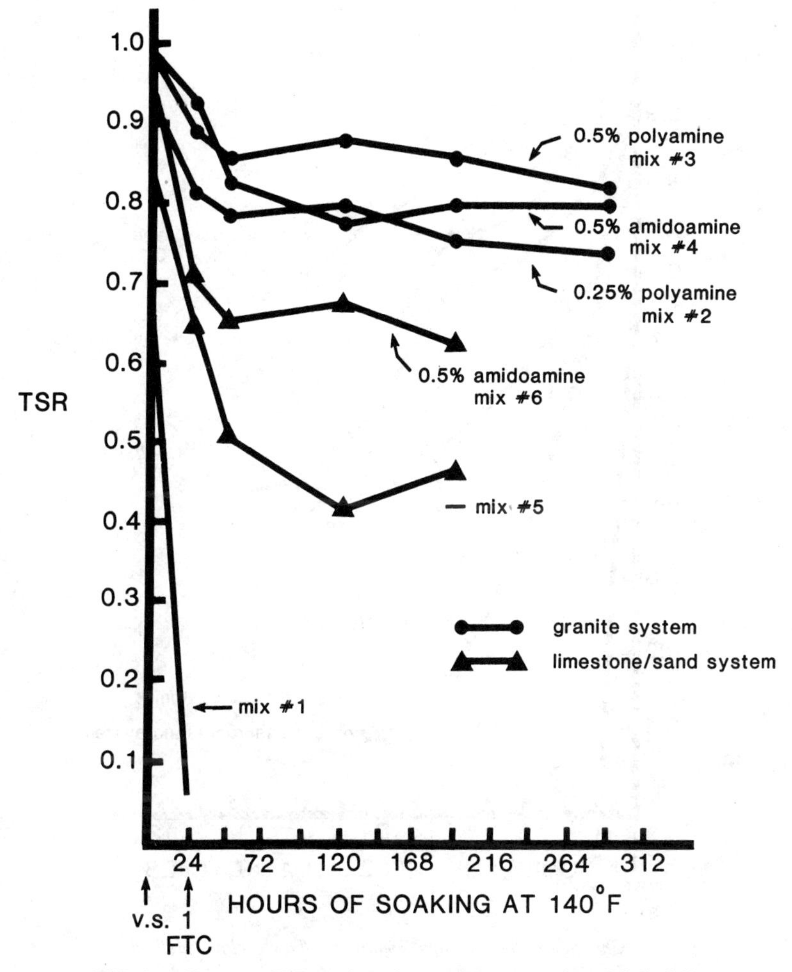

FIG. 1—*Change in TSR resulting from continuous soaking (Method A).*

after the first FTC, and this damage accumulates progressively with each additional FTC. The ranking of mixture performance on a final TSR basis is the same for both Methods A and B and this similarity points out the general and unbiased nature of the additional moisture damage produced by Method B. Additionally, it is evident that changes in additive concentration that produce minor TSR differences after 1 FTC become more significant as the FTC process is extended (compare Mixtures 2 and 3).

Data for Mix 3 were used in generating the comparison of Methods A and B shown in Fig. 3. The TSR drop as a function of soak time at 60°C (140°F) clearly demonstrates that the freeze portions of Method B catalyze additional moisture damage. In fact, the rate of TSR drop is nearly an order of magnitude greater for

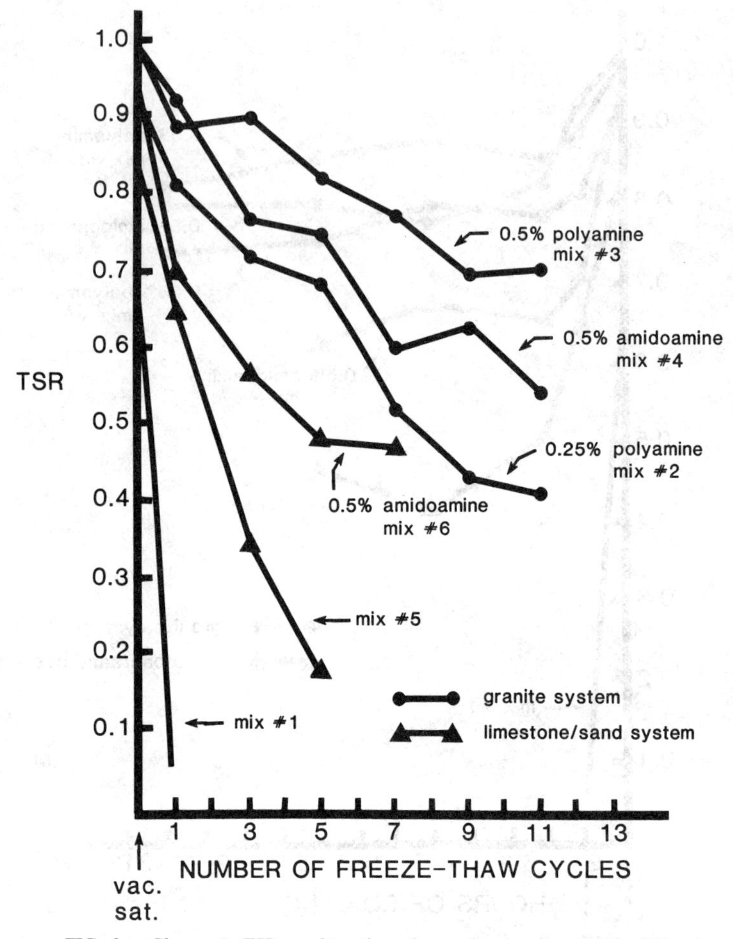

FIG. 2—*Change in TSR resulting from freeze-thaw cycling (Method B).*

the repetitive freeze-thaw cycle procedure as compared to continuous soaking. This rate difference is potentially a significant factor in asphalt concrete durability. Therefore, we chose to adopt Method B as a general procedure in a more detailed investigation of moisture susceptibility.

Method Utilization

From Fig. 2 it is evident that the 0.5% polyamine treatment reduces the rate of TSR loss for the granite asphalt concrete system by approximately twenty-fold when compared to no treatment. This reduction in moisture damage rate is significant, but damage elimination is ideally more desirable. In order to more thoroughly examine the mixture factors that influence the moisture damage process, a series of related asphalt concrete specimens were prepared and evaluated.

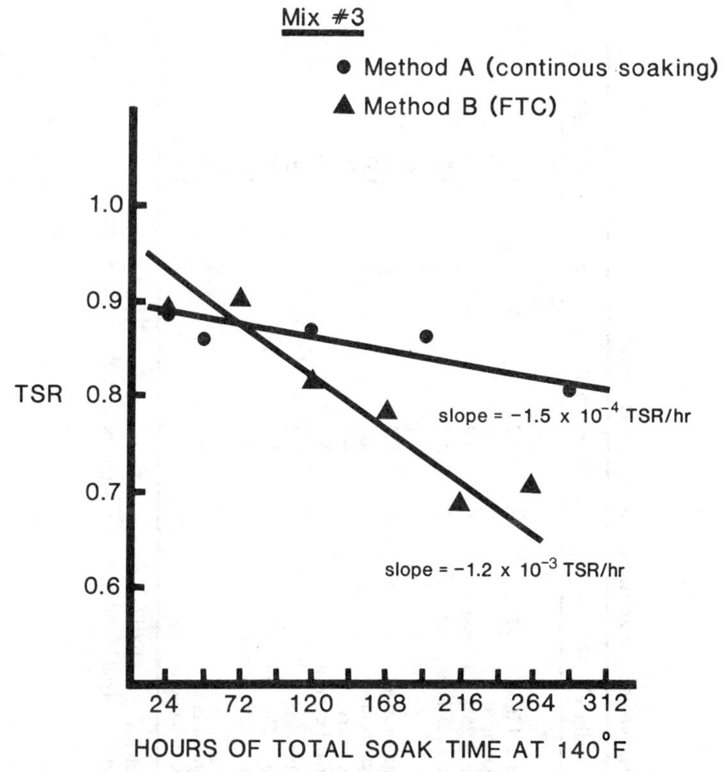

FIG. 3—*Comparison of rate of TSR changes for Methods A and B.*

Three major material variables were examined and the specimens from each aggregate, asphalt, and additive are fully described in Table 2.

The changes in TSR values with increasing FTL exposure for Mixes 7 through 22 (Table 3) permit certain observations regarding material effect on damage rate. The viscosity grade of asphalt cement does not significantly change the performance of an aggregate that is extremely moisture sensitive (Mixes 12 and 20). Aggregates within the same region and same general mineralogical classification may, however, have appreciable differences in FTC damage rates. For instance, the Kennesaw aggregate system (Mix 7) loses tensile strength six times slower than the Lithonia aggregate system (Mix 12). Despite improvements that arise from changing aggregates, the best deterrent to TSR loss is provided by additive treatments.

The extent to which additives alter the performance and durability of the granite gneiss asphalt concrete mixtures is significant in all cases and is easily recognized. Differentiation among the specific additives is somewhat more complex and must involve consideration of absolute TSR values as well as rates of TSR loss (Table 4). On this basis, the repetitive FTC method does not detect substantial performance differences between hydrated lime added dry to the

TABLE 2—*Data for Lithonia and Kennesaw granite gneiss asphalt concrete mixtures.*

Mix Number	Aggregate	Asphalt[a]	Additive[b]	Percent Air Voids	Dry Tensile Strength, psi
7	Kennesaw	AC-20	none	5.8	124
8	Kennesaw	AC-20	1.0% hydrated lime	5.2	120
9	Kennesaw	AC-20	0.25% polyamine	6.2	126
10	Kennesaw	AC-20	1.0% lime/0.25% polyamine	5.4	125
11	Kennesaw	AC-20	0.5% lime/0.25% polyamine	5.4	129
12	Lithonia	AC-20	none	6.7	127
13	Lithonia	AC-20	1.0% hydrated lime	5.4	139
14	Lithonia	AC-20	0.5% hydrated lime	5.7	137
15	Lithonia	AC-20	1.0% lime slurry	6.2	130
16	Lithonia	AC-20	1.0% portland cement	5.6	134
17	Lithonia	AC-20	0.25% polyamine	6.2	129
18	Lithonia	AC-20	1.0% lime/0.25% polyamine	5.6	129
19	Lithonia	AC-20	0.5% lime/0.25% polyamine	5.8	134
20	Lithonia	AC-30	none	6.9	108
21	Lithonia	AC-30	1.0% hydrated lime	6.0	107
22	Lithonia	AC-30	0.25% polyamine	6.9	110

[a]The AC-20 and AC-30 asphalt cements were included as 5.5 weight percent of the total mixture.
[b]The mineral additives are included as a weight percent of aggregate and the polyamine additive as a weight percent of asphalt cement.

TABLE 3—*Freeze-thaw cycle performance for Lithonia and Kennesaw granite gneiss asphalt concrete mixtures.*

Mix Number	Additive	Tensile Strength Ratio				
		1 FTC	3 FTC	5 FTC	7 FTC	9 FTC
7	none	0.38	0.25	0.00	...	...
8	1.0% hydrated lime	1.07	0.98	0.92	0.85	0.76
9	0.25% polyamine	0.72	0.63	0.48	0.41	0.43
10	1.0% lime/0.25% polyamine	1.10	1.05	1.00	0.95	0.91
11	0.5% lime/0.25% polyamine	1.11	1.08	1.04	1.01	0.99
12	none	0.00	...	...	...	...
13	1.0% hydrated lime	0.99	0.95	0.86	0.83	0.78
14	0.5% hydrated lime	1.04	1.05	0.83	0.78	0.72
15	1.0% lime slurry	1.10	0.97	0.97	0.91	0.89
16	1.0% portland cement	0.96	0.87	0.86	0.80	0.77
17	0.25% polyamine	0.69	0.52	0.45	0.40	0.35
18	1.0% lime/0.25% polyamine	1.09	1.01	0.98	0.94	0.94
19	0.5% lime/0.25% polyamine	1.07	0.97	0.95	0.95	0.96
20	none	0.00	...	...	...	...
21	1.0% hydrated lime	0.96	0.92	0.80	0.79	0.77
22	0.25% polyamine	0.78	0.69	0.61	0.51	0.46

TABLE 4—*Influence of additive on moisture damage rates for Lithonia granite gneiss specimens.*

Mix Number	Approximate Damage Rate,[a] TSR/h
12	-3×10^{-2} (minimum)
13	-1×10^{-3}
14	-2×10^{-3}
15	-1×10^{-3}
16	-1×10^{-3}
17	-2×10^{-3}
18	-7×10^{-4}
19	-4×10^{-4}

[a]Determined as in Fig. 3.

aggregate, lime slurry, or portland cement. This suggests that the mineral additives or addition methods are interchangeable for imparting improved moisture resistance to the Lithonia aggregate.

Performance deviations from lime and portland cement occur with the polyamine treatment of asphalt cement or the combination of polyamine and lime treatments of the asphalt cement and aggregate, respectively. The combination treatment is especially effective as very high TSR values are achieved, and the TSR loss rate is nearly two orders of magnitude lower than the no additive TSR loss rate. The exact reasons for the surprising effectiveness are not known, but it is clear that the combination treatment nearly provides the ideal model for an asphalt concrete system which is durable and insensitive to potential moisture damage.

Certain parameters in the chemical makeup of Mixtures 13 and 17 were measured in order to detect composition changes during the repeated cycling process. The amount of free polyamine in specimens from Mixture 17 was determined by reaction with a diazonium salt, and the lime content in Mixture 13 was monitored through surface analysis for calcium by XRF. Surface iron on the Lithonia aggregate served as an internal quantitative reference in the XRF method.

The polyamine and calcium (lime) contents both fall as a result of the moisture conditioning. The relationship between these ingredient changes and the variations in TSR is provided in Fig. 4. The "parallelism" of TSR and ingredient content changes suggests that physiochemical changes at the aggregate surface or within the asphalt binder are associated with moisture damage, and this damage cannot be solely attributed to mechanical forces that arise in the FTC process.

Method Uses in Product Development

Because of the superior performance properties of lime/polyamine combination treatments some attention was given to introducing these properties into a single additive. Simple mixtures of lime or other calcium salts with polyamines and amidoamines did not provide the desired performance when added either to the aggregate or to the asphalt cement. The most promising results were discovered with experimental amine complexes of certain transition metals. Two com-

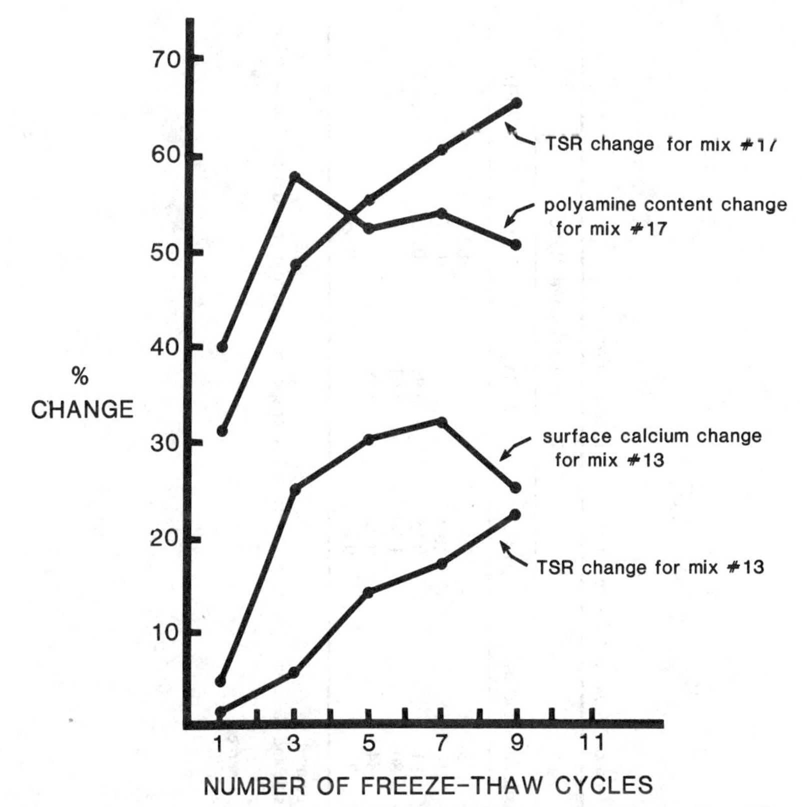

FIG. 4—*TSR and chemical composition changes for asphalt concrete Mixtures 13 and 17.*

plexes, A and B differing in the transition metal, were examined extensively by conventional tensile strength, resilient modulus, and fatigue life tests. The results from these tests are summarized in Table 5.

Comparison of the performance for Complexes A and B does not indicate a significant advantage for either material. Repetitive FTC studies, however,

TABLE 5—*Indirect tensile properties for wet and dry asphalt concrete mixtures containing metalloamine complexes.*

Additive[a]	Tensile Strength Ratio[b]	$M_R R$[c]	Fatigue Life Ratio[d]
1.0% hydrated lime[e]	0.99	0.72	1.67
0.5% Complex A[e]	1.07	0.95	1.36
0.5% Complex B[e]	0.99	0.85	1.51

[a]Lime content is based on aggregate weight and complex content on asphalt weight.
[b]One FTC only.
[c]Expressed as one FTC resilient modulus divided by dry resilient modulus.
[d]Related to the maximum field bending stress that produces fatigue failure at 100 000 load cycles. Expressed as 1 FTC stress divided by dry stress.
[e]Specimens derived from Lithonia granite gneiss and AC-20 asphalt cement.

TABLE 6—*Results of asphalt concrete materials obtained during Virginia test section construction.*

Additive[a]	Percent Voids[b]	Dry Tensile Strength, psi	Tensile Strength Ratio			
			1 FTC	3 FTC	5 FTC	7 FTC
0.7% BA-2000[c]	4.1	135	0.98	...	0.99	...
0.7% BA-2000[c]	5.7	110	1.01	...	0.97	...
0.7% BA-2000[c]	7.0	91	0.98	0.96	0.90	0.85
0.7% BA-2000[d]	6.8	80	1.00	0.91	0.78	0.70
none[d]	6.7	84	0.32	0.19	...	...
1.0% hydrated lime[d]	6.2	72	0.82	0.71	0.65	0.53

[a]All specimens prepared with granite aggregate and AC-20 asphalt cement.
[b]The specimens in the first two listings were prepared with 50 blows/side and 30 blows/side, respectively. All other specimens were prepared with 15 blows/side.
[c]From batch plant production.
[d]Specimens prepared with laboratory mixing of aggregate and asphalt cement.

clearly reveal that Complex A is the superior additive as it provides a TSR loss rate of -1×10^{-3} TSR/h compared to -3×10^{-3} TSR/h for Complex B. Since all other performance properties are so similar, the TSR loss rate becomes the decisive factor in determining which additive exhibits the better potential for long-term field durability.

Further evaluations of Complex A (commercially known as CARSTAB BA-2000) and the predictive value of the repetitive FTC method were initiated with the construction of pavement test sections. Several state highway departments are kindly cooperating in this ongoing investigation, and an example (Virginia) of relevant test section data is presented below.

Asphalt cement, aggregate, and hot-mix asphalt concrete were collected during test section construction on U.S. Highway 360 near Burkeville, VA. These materials were used to prepare a series of laboratory-tested specimens that matched pavement density (7% voids), varied additive treatments, and compared batch plant generated mixtures to laboratory mixtures. The repetitive FTC test results for these specimens are presented in Table 6.

Here again, Complex A considerably reduces the FTC damage that occurs in the untreated asphalt concrete specimens. There appears to be a subtle performance difference between specimens prepared with batch plant mixing compared to those prepared with laboratory mixing. The batch plant specimens (at equal void levels) are slightly stronger and more durable to the repetitive FTC conditioning process. The additional heat history or laboratory storage experienced by the batch plant specimens may account for these differences [*3*]. Despite this difference, the effectiveness predicted for the Complex A additive by FTC conditioning is not greatly altered by field production factors that are not easily simulated in the laboratory.

The fatigue life properties of specimens corresponding to the third entry in Table 6 were also evaluated in order to determine if unexpected life changes occur as a result of the repetitive FTC conditioning. Illustrations representing deformation rates and fatigue lines for these specimens are presented in Figs. 5 and 6, respectively. The stress fatigue life ratio at each FTC (determined as described in Table 4) is greater than the corresponding TSR value (TSR = 0.80 for 9 FTC). Such a relative positioning of the two ratio values is consistent with previous reports for mixtures exposed only to 1 FTC [*6*], and it is evident that repetitive FTC exposure does not significantly alter this relationship.

Conclusion

The two extended water exposure methods used in this research to probe the long-term durability of additives and asphalt concrete accelerate the moisture damage process at significantly different rates. The harshest conditioning is provided by the repetitive freeze plus warm-water soak procedure, and the TSR and fatigue life studies demonstrate the freezing portion of this method to be a catalyst for initiating further damage. This damage is not solely mechanical in

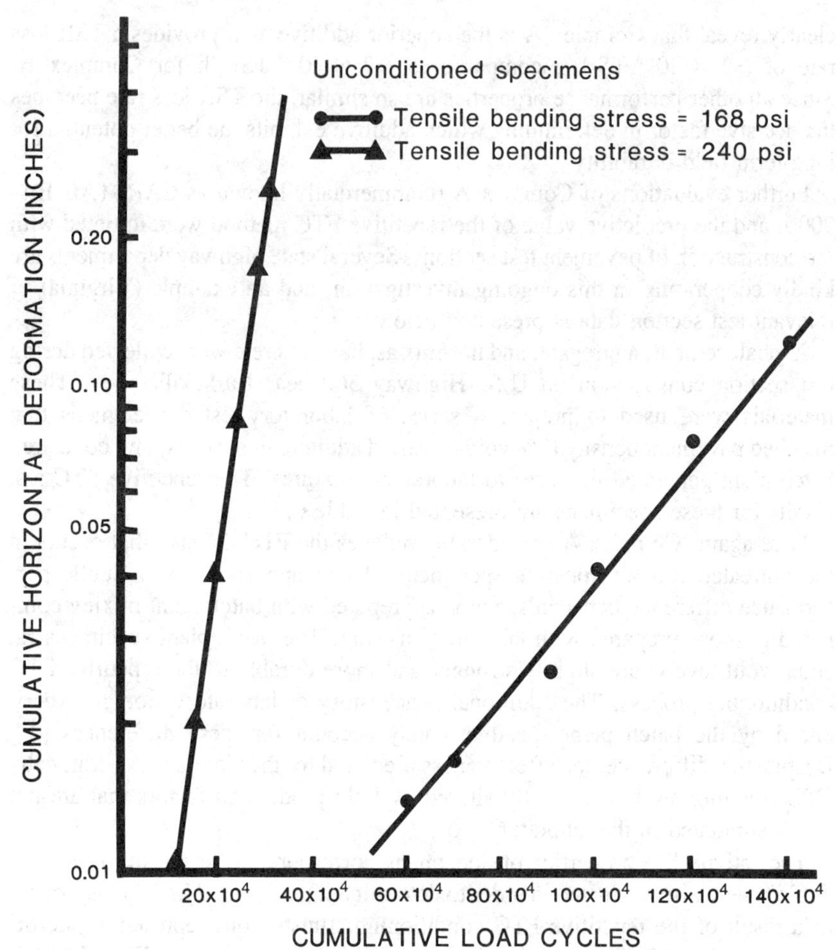

FIG. 5—*Fatigue deformation rates for Virginia plant mix asphalt concrete specimens containing BA-2000.*

nature as compositional analyses of cyclic specimens reveal that chemical changes occur simultaneously with damage accumulation.

The rate at which moisture damage accumulates during the repetitive FTC process is influenced by several factors. The most prominent factor found in this investigation is the change in chemical or physiochemical properties of an asphalt concrete mixture because of additive treatment. For example, certain additive treatments reduce the TSR loss rate by nearly two orders of magnitude when compared to no treatment. A comparison of TSR loss rates is also informative on the relative effectiveness of additive treatments and provides important assistance in the development of additives with improved long-term durabilities.

The extended exposure procedures described here are not practical for routine use in most laboratories, and the conditions themselves may be more severe than

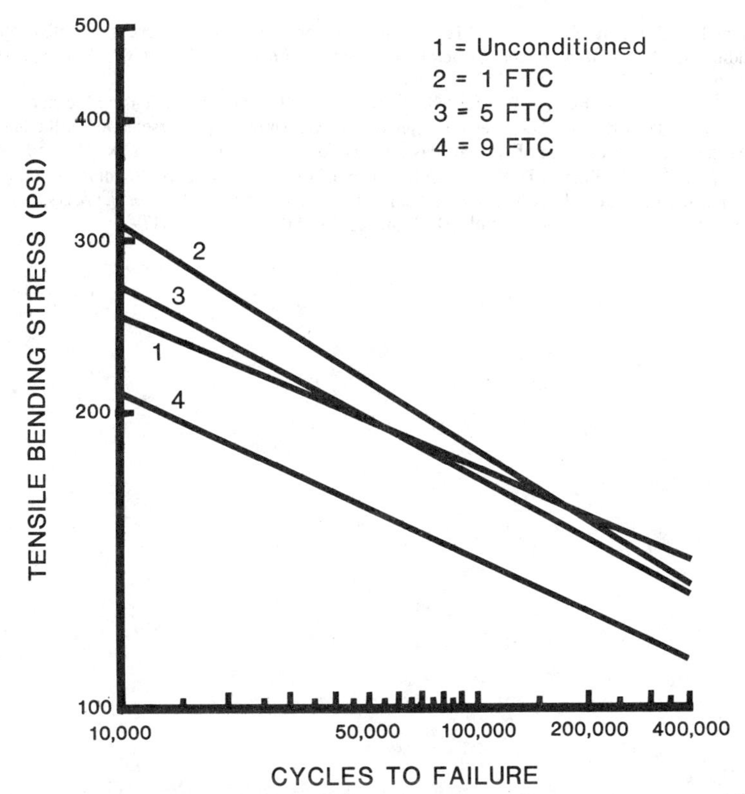

FIG. 6—*Stress fatigue lines for Virginia plant mix specimens after exposure to repeated freeze-thaw cycles.*

those experienced by pavements in many sections of the world. Nevertheless, some estimation of moisture damage rate with an appropriate conditioning process should be obtained for asphalt concrete materials used in pavement construction. The simplest technique for obtaining such rate information is to measure the tensile strength of dry, vacuum saturated, and 1 FTC conditioned specimen. If concern over the TSR loss rate is generated with this minimal information, then additional moisture conditioning is recommended for problem verification and alternative material selection.

References

[*1*] Taylor, M. A. and Khosla, N. P., "Stripping of Pavements: State of the Art," *Transportation Research Record 911,* Washington, DC, 1983, pp. 150–157.

[*2*] Lottman, R. P., "Predicting Moisture-Induced Damage to Asphaltic Concrete," *National Cooperative Highway Research Program Report 192,* Washington, DC, 1978.

[*3*] Lottman, R. P., "Predicting Moisture-Induced Damage to Asphaltic Concrete Field Evaluation Phase," *National Cooperative Highway Research Program Report 246,* Washington, DC, 1982.

[*4*] Abel, F. and Proctor, J., "High Altitude Premium Pavements," *Report CHOD-SMB-R-80-4,* Colorado Department of Highways, 1980.

[5] Tunnicliff, D. G. and Root, R. E., "Testing Asphalt Concrete for Effectiveness of Antistripping Additives," *Proceedings of the Association of Asphalt Paving Technologists,* Vol. 52, Minneapolis, MN, 1983, pp. 535–553.

[6] Gilmore, D. W., Lottman, R. P., and Scherocman, J. A., "Use of Indirect Tension Measurements to Examine the Effect of Additives on Asphalt Concrete Durability," presentation at the annual meeting of the Association of Asphalt Paving Technologists, Minneapolis, MN, 11 April 1984.

[7] Dalter, R. S. and Gilmore, D. W., "A Comparison of Effects of Water on Bonding Strengths of Compacted Mixtures of Treated Versus Untreated Asphalt," *Proceedings of the Association of Asphalt Paving Technologists,* Vol. 51, Minneapolis, MN, 1982, pp. 317–326.

Mang Tia[1] *and Leonard E. Wood*[2]

The Use of Water Immersion Tests in the Evaluation of the Effects of Water on Cold-Recycled Asphalt Mixtures

REFERENCE: Tia, M. and Wood, L. E., **"The Use of Water Immersion Tests in the Evaluation of the Effects of Water on Cold-Recycled Asphalt Mixtures,"** *Evaluation and Prevention of Water Damage to Asphalt Materials, ASTM STP 899,* B. E. Ruth, Ed., American Society for Testing and Materials, Philadelphia, 1985, pp. 89–103.

ABSTRACT: The effectiveness of various immersion-mechanical tests in measuring the water susceptibility of cold-recycled asphalt mixtures was investigated. The immersion-mechanical tests involved subjecting compacted recycled asphalt mixtures to a 24-h water immersion process followed by mechanical tests. The various mechanical tests used in this study included the resilient modulus, Marshall stability, Hveem stabilometer *S* value, Hveem stabilometer *R* value, and Hveem cohesiometer tests. The test values obtained from the wet specimens were compared to the values obtained from the dry specimens, and the effectiveness of the test methods was evaluated.

The results of the study indicated that the resilient modulus, Marshall stability, Hveem *S* value, and Hveem cohesiometer tests were effective in measuring the changes in stiffness, stability, and cohesion of cold-recycled mixtures caused by water damage. The Hveem *R*-value test was not very sensitive in measuring the effect of water when a recycled mixture was relatively stable. However, the *R*-value test became very sensitive when a mixture was unstable, and thus it was very effective in determining the recycled mixtures that were highly water-susceptible. The water resistance of cold-recycled mixtures generally increased with curing time and compactive effort. The effects of the type of added virgin binder and the type of added virgin aggregate to the water resistance of cold-recycled asphalt mixtures were significant. The proper consideration of curing time, compactive effort, added virgin binder, and added virgin aggregate will help to attenuate the effect of water on cold recycled asphalt mixtures.

KEY WORDS: pavements, asphalts, cold-recycled asphalt mixtures, immersion mechanical tests, resilient modulus, Hveem *S* value, Hveem *R* value, cohesiometer, Marshall stability, curing time, compactive effort, virgin binder, virgin aggregate

[1]Assistant professor, Department of Civil Engineering, Weil Hall, University of Florida, Gainesville, FL 32611.

[2]Professor, School of Civil Engineering, Civil Engineering Building, Purdue University, West Lafayette, IN 47907.

Asphalt pavement recycling basically involves the addition of virgin aggregates to upgrade the old aggregate and the addition of rejuvenating agents or soft asphalts to restore the old hardened asphaltic binders to their desirable properties. It can be done by either a hot or a cold process. As compared to the hot-recycled asphalt mixtures, the cold-recycled asphalt mixtures tend to have a less homogeneous coating of virgin asphalts or rejuvenating agents during the recycling process. Thus, the bonding between the bitumen-aggregate pieces in the cold-recycled asphalt mixtures may be poorer, and the mixtures may be more susceptible to water damage. In the design of cold-recycled asphalt mixtures, suitable tests should be used to evaluate the water resistance of these mixtures, so that the recycled pavements can be properly designed and water damage to the pavement structures can be alleviated.

The water damage of cold-recycled asphalt mixtures may be manifested in one or combinations of the following forms of deterioration:

1. The separation of bitumen-aggregate pieces caused by the loss of adhesion between the added binders or rejuvenating agents and the bitumen-aggregate pieces.
2. The reduction in the strength and the stability (resistance to permanent deformation) of the asphalt mixtures.
3. The reduction in the stiffness of the asphalt mixtures.

These moisture-induced damages may be caused by (1) water saturation, (2) freezing and thawing, or (3) washing action. Tests that evaluate the water resistance of asphalt mixtures generally measure the changes in physical properties after the mixtures have been subjected to some simulated actions of water. The three general categories of water resistance tests classified according to the simulated actions of water are

1. *Immersion-Mechanical Tests*—Asphalt mixture specimens are soaked in water for a specified period of time, and their changes in physical properties after the soaking process are measured. The mechanical tests used include the resilient modulus tests [*1*,*2*], the Marshall stability test [*3*,*4*], the indirect tension test [*5*,*6*], the Hveem *S*-value test [*7*], the Hveem *R*-value test [*1*,*7*], and the ASTM Test for Effect of Water on Cohesion of Compacted Bituminous Mixtures (D 1075).
2. *Freezing-Thawing Tests*—An example of freezing-thawing tests is the Texas freeze-thaw pedestal test, which measures the number of freeze-thaw cycles that an asphalt specimen can endure before cracking, and is used as a measure of water susceptibility [*5*,*8*].
3. *Boiling Tests*—Asphalt mixtures are subjected to the action of boiling water for a specified time, and the extent of stripping of binders from the aggregate is evaluated. Examples of boiling tests include the ASTM Test for Effect of Water on Bituminous-Coated Aggregate-Quick Field Test (D 3625) and the Texas Boiling Test [*5*,*8*].

The last two categories of tests, the freezing-thawing and the boiling tests, have generally been used for identifying those asphalt mixtures that have stripping problems. Results from these tests can be used as a guide for acceptance or rejection of some asphalt mixtures. The immersion-mechanical tests have generally been used for determining the loss in strength and stiffness of mixtures resulting from exposure to moisture. The reduced strength and stiffness values obtained from these tests can be used in pavement and mix designs. In this study, the effectiveness of various immersion-mechanical tests in measuring the water susceptibility of cold-recycled asphalt mixtures was investigated. The various mechanical tests used in this study include the resilient modulus, the Marshall, the Hveem stabilometer *S* value, the Hveem stabilometer *R* value, and the Hveem cohesiometer tests.

Test Methods

Water Immersion Procedure

The water sensitivity test as recommended by the Asphalt Institute [*9*] was slightly modified and used in this study. The test specimen is first subjected to a vacuum of 30-mm mercury for 1 h. After the 1-h period, water at room temperature of 22°C is drawn into the vacuum chamber, submerging and vacuum saturating the specimen. The vacuum is then released, and the specimen is soaked in the water bath for 24 h before testing. This water-immersion procedure simulates the action on water saturation on an asphalt mixture. The modification from the Asphalt Institute procedure was that water at room temperature of 22°C, rather than the standard 25°C, was used.

Diametral Resilient Modulus Test

Resilient modulus is defined as the ratio of the applied stress to the recoverable strain when a repeated dynamic load is applied. In the diametral resilient modulus test, dynamic pulse loads are applied diametrally to Marshall size specimens, and the induced vertical or horizontal deformations or both are recorded and used to calculate the resilient moduli. The diametral resilient modulus tests in this study was modified from the one proposed by Schmidt [*10*], in that the vertical deformation rather than the horizontal deformation was used to calculate the resilient modulus of the test specimen. The resilient modulus can be calculated from the following relationship [*1*]

$$M_R = 3.583P/td_v$$

where

M_R = resilient modulus,
P = applied load,
t = thickness of test specimen, and
d_v = recoverable vertical deformation.

This relationship holds for a Marshall-size specimen, (101.6 mm [4 in.] in diameter) loaded diametrally.

Marshall Stability Test

Marshall stability tests were run in accordance with the ASTM Test Method for Resistance to Plastic Flow of Bituminous Mixtures Using Marshall Apparatus (D 1559) with the exception that tests were run at room temperature (22°C) rather than the standard temperature of 60°C.

Hveem Stabilometer S-*Value Test*

Hveem stabilometer *S*-value tests were run in accordance with the ASTM Test Methods for Resistance to Deformation and Cohesion of Bituminous Mixtures by Means of Hveem Apparatus (D 1560), with the exception that tests were run at room temperature (22°C) rather than the standard temperature of 60°C.

Hveem Stabilometer R-*Value Test*

Hveem stabilometer *R*-value tests were run in accordance with the ASTM Test Method for Resistance *R*-Value and Expansion Pressure of Compacted Soil (D 2844).

Hveem Cohesiometer Test

Hveem cohesiometer tests were run in accordance with the procedures of ASTM D 1560.

Cold-Recycled Asphalt Mixtures

Old Pavement Materials

Two old pavement materials, obtained from two state roads in Indiana, were used to make the cold-recycled mixtures for this study. The pavement materials were crushed to a maximum size of 2.54 cm (1 in.) and sieved into various size groups. For each batch of recycled mixture, the pavement materials in all size groups were recombined in the same proportion to obtain homogeneity in the mixes.

Pavement Material 1 had an average bitumen content of 5.0% by weight of aggregate. Its recovered bitumen had a penetration of 25 (dmm) at standard condition (100 g, 5 s, 25°C) and an absolute viscosity of 6380 Pa·S (63 800 poises) at 60°C. The recovered aggregate consisted mainly of crushed limestone. The gradation of the recovered aggregate is shown in Fig. 1. It is compared to the gradation range of Indiana's Type II No. 9 surface mix aggregate.

Pavement Material 2 had a bitumen content of 4.5% by weight of aggregate. The recovered bitumen had a penetration of 38 at standard condition (100 g, 5 s,

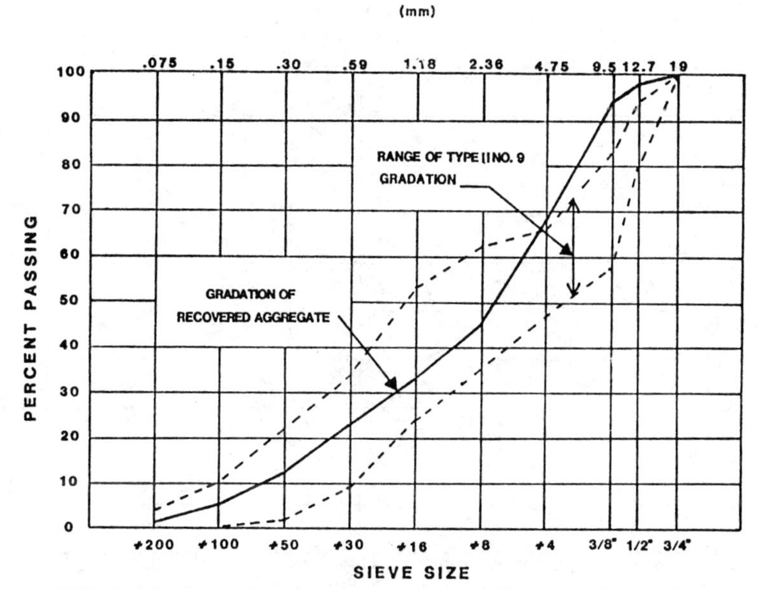

FIG. 1—*Gradation of recovered aggregate from old Pavement Material 1.*

25°C). The recovered aggregate contained mainly sand and gravel. The gradation of the recovered aggregate is depicted in Fig. 2. It is compared to the gradation range of Indiana's Type II No. 8 surface mix aggregate.

Added Binders

The three asphaltic binders that were added to the old pavement materials to make the recycled mixes in this study were (1) a high-float mixing grade asphalt emulsion designated AE-150, (2) a high-float mixing grade asphalt emulsion designated AE-90, and (3) a foamed asphalt made from a soft asphalt designated AC-2.5. The physical properties of AE-150, AE-90, and AC-2.5 are described in Tables 1 to 3, respectively.

Specimen Preparation Procedure

The cold-recycled asphalt mixtures used for this study were prepared in the laboratory. The specimen preparation procedure consisted of the following general steps:

1. The proper amount of old pavement material was batched for one specimen.
2. The required amount of water was added to the material and mixed thoroughly with a mechanical mixer and then with a spoon by hand. The material was then left for 10 to 15 min.
3. The proper amount of binder was added to the material and mixed with a mechanical mixer for 1.5 min and with a spoon for 30 s.

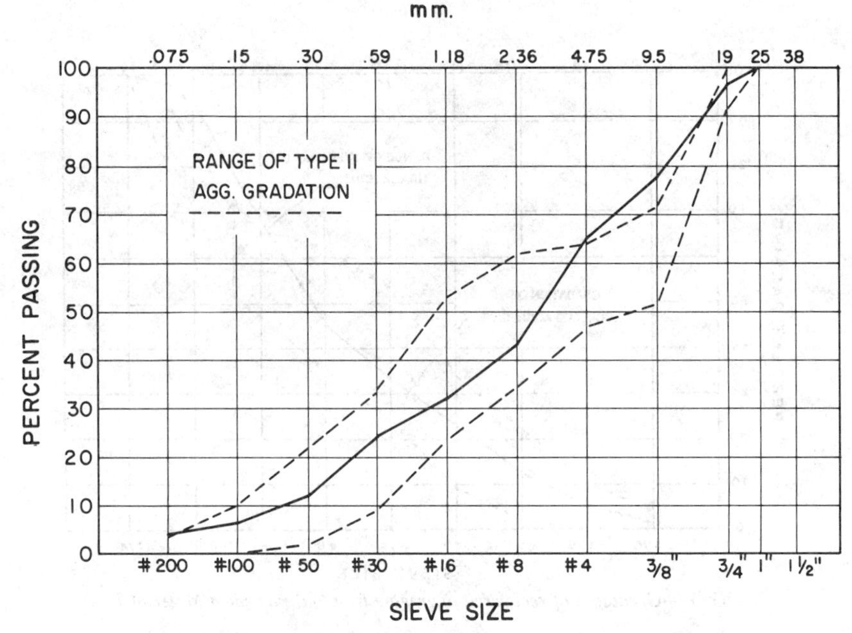

FIG. 2—*Gradation of recovered aggregate from old Pavement 2.*

4. The mix was cured for 1 h in a forced-draft oven at 60°C (140°F).
5. The mix was remixed for 30 s with a mechanical mixer and was compacted immediately in the gyratory machine.
6. After compaction, the specimen was extruded from the mold within 30 min and left to cure at room temperature.

Ultimate Curing Condition

To simulate the "ultimate" condition of the compacted recycled mixtures after a long curing period, specimens were placed in a forced draft oven at 60°C for 24 h. The recycled mixtures at "ultimate" curing condition had zero moisture content.

Mixing Water

The purpose of adding water to the mix was to facilitate the mixing process. When asphalt emulsion was used as the added binder, 1% water was added. When foamed asphalt was used, 3% water was added.

Gyratory Compaction

The standard gyratory compaction procedure as specified in ASTM D 3387 was generally followed. The initial gyratory angle was set at 1°, and a fixed roller was used. The ram pressure was set at 1.38 MPa (200 psi). Various levels of

TABLE 1—*Physical properties of AE-150.*

Property	Standard	Test Condition	Value
Percent residue by distillation	ASTM D 244	standard	70.0%
Oil portion of distillate	ASTM D 244	standard	1.5%
	TEST ON DISTILLATION RESIDUE		
Penetration	ASTM D 5	100 g, 5 s, 25°C	215 (dmm)
Specific gravity	ASTM D 70	25°C	1.010
Float	ASTM D 139	60°C	>200 s

TABLE 2—*Physical properties of AE-90.*

Property	Standard	Test Condition	Value
Percent residue by distillation	ASTM D 244	standard	71.0%
Oil portion of distillate	ASTM D 244	standard	0%
	TEST ON DISTILLATION RESIDUE		
Penetration	ASTM D 5	100 g, 5 s, 25°C	120 (dmm)
Specific gravity	ASTM D 70	25°C	1.010
Float	ASTM D 139	60°C	200 s

TABLE 3—*Physical properties of AC-2.5.*

Property	Standard	Test Condition	Value
Penetration	ASTM D 5	100 g, 5 s, 25°C	>300 (dmm)
Absolute viscosity	ASTM D 2170	60°C	300 poise
Kinematic viscosity	ASTM D 2170	135°C	160 cSt
Specific gravity	ASTM D 70	25°C	1.024
Ductility	ASTM D 113	25°C	>100 cm

compactive effort were obtained by varying the number of revolutions of the gyratory machine.

Added Virgin Aggregate

The virgin aggregate used in this study was a sand and gravel from the Western Indiana Aggregate Corporation gravel pit in West Lafayette, IN. It consists mainly of limestone, dolomite, and a small amount of granite and quartz.

Results

Resilient Modulus Test Results

Figure 3 presents the resilient moduli of the recycled mixtures with AE-150 as the added binder under the wet (after the 24-h water-immersion procedure) and the dry conditions. These recycled mixtures were made from old Pavement Material 1, described in the previous section. It is clear from the figure that the

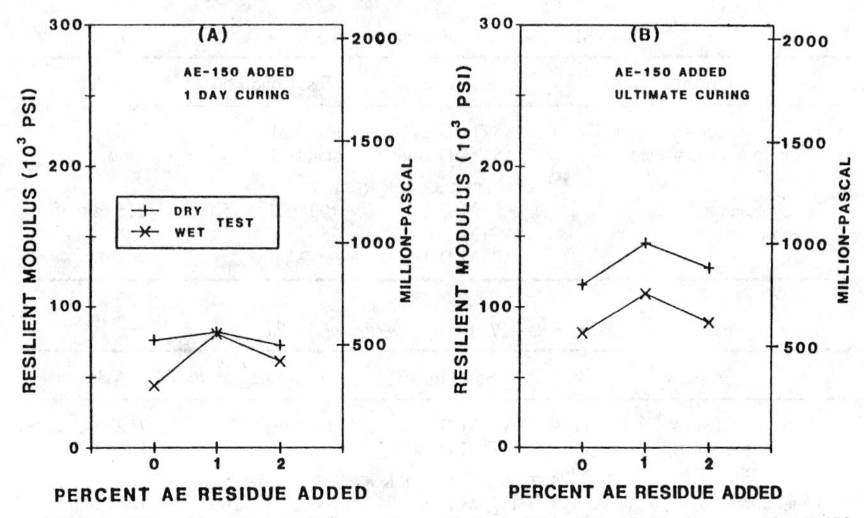

FIG. 3—*Effect of water immersion on the resilient modulus of recycled mixtures with AE-150 added.*

optimum AE residue added is 1% for both the dry and the wet conditions. The resilient moduli at the wet condition are significantly lower than those at the dry condition. It can also be noted that the "wet" resilient moduli at the ultimate curing are significantly higher than those at one-day curing.

Marshall Stability Test Results

The Marshall stabilities of the same recycled mixtures (made from old pavement material 1, with AE-150 and foamed asphalts as the added binders) under the dry and the wet conditions are presented in Fig. 4. The reduction in stability

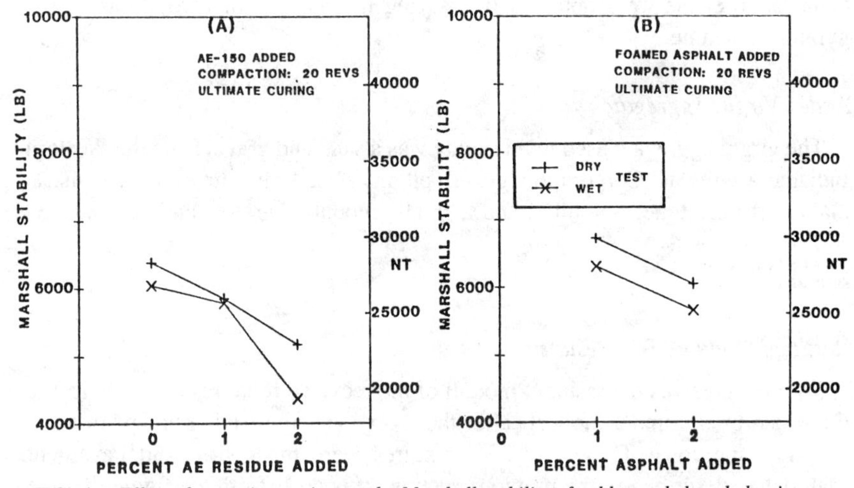

FIG. 4—*Effect of water immersion on the Marshall stability of cold-recycled asphalt mixtures.*

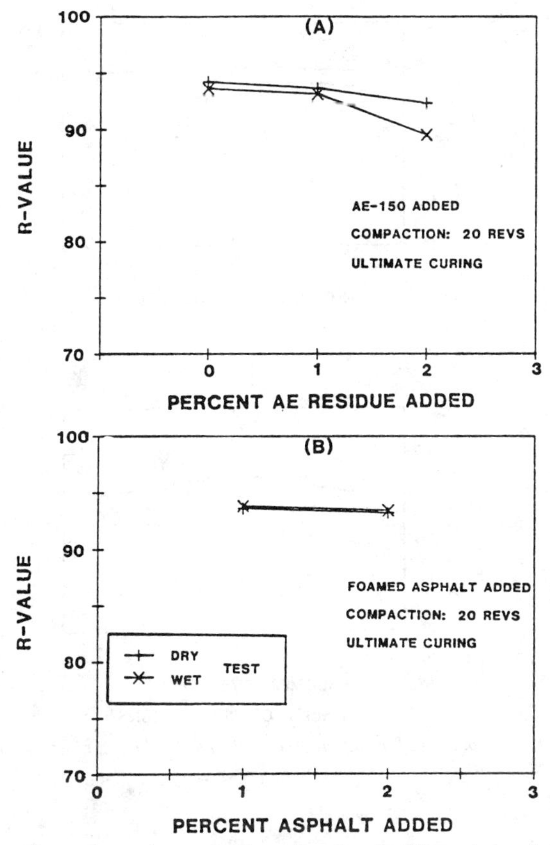

FIG. 5—*Effect of water immersion on the Hveem* R*-value of cold-recycled asphalt mixtures (made from Pavement Material 1).*

of the recycled mixtures caused by the action of water is clearly shown here. It is evident that the immersion Marshall stability test is sensitive in measuring the effect of water on these mixtures. It can also be observed that the recycled mixtures with foamed asphalt added have higher stability values under both the dry and the wet conditions.

Hveem R*-Value Test Results*

The Hveem R-values of the same recycled mixtures (made from old Pavement Material 1, with AE-150 and foamed asphalt as the added binders) under the dry and the wet conditions are presented in Fig. 5. It can be noted that the R value test is not very sensitive in measuring the effect of water on a recycled mixture when the retained stability of the recycled mixture is relatively high. However, the R-value test becomes very sensitive when a mixture is unstable, as in the case of the mixture with 2% AE residue added. This general conclusion is supported by other test results in this study.

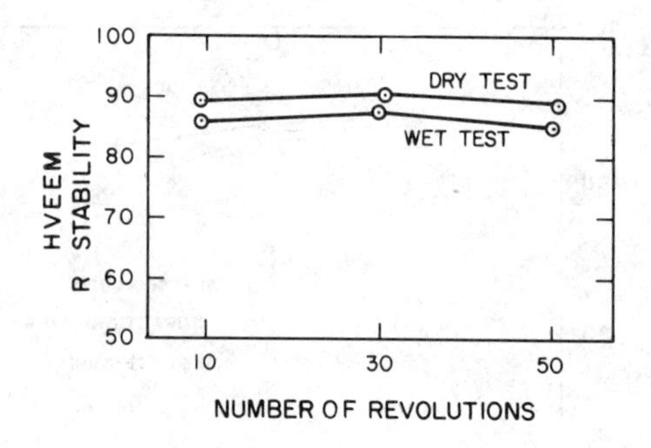

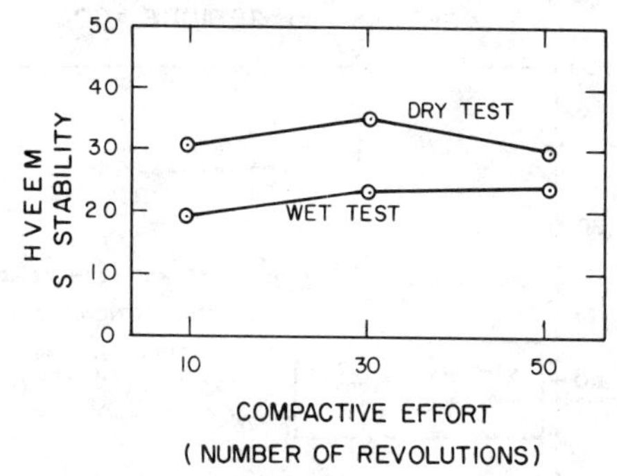

FIG. 6—*Effect of compactive effort on the Hveem* R *and* S *values of recycled mixtures under dry and wet conditions.*

Hveem S*-Value Test Results*

Significant reduction in *S* value caused by effect of water immersion was noted among the recycled mixtures that had been indicated to be water-susceptible by the resilient modulus and the Marshall stability tests. The immersion *S*-value test is effective in determining the effect of water on these recycled mixtures.

Cohesiometer Test Results

Cold-recycled asphalt mixtures show a reduction in cohesion when soaked in water. The reduction in cohesion of a mixture caused by water damage can be quantified by cohesiometer tests under the dry and the wet conditions. Significant reduction in cohesiometer value caused by the effect of water immersion was noted among all the recycled mixtures tested. The immersion cohesiometer test is effective in measuring the changes in cohesion of the recycled mixtures caused by the action of water. However, it should be noted that the cohesiometer test has to be run at 60°C (104°F). At this elevated temperature, the properties of a cold-recycled asphalt mixture may be slightly changed.

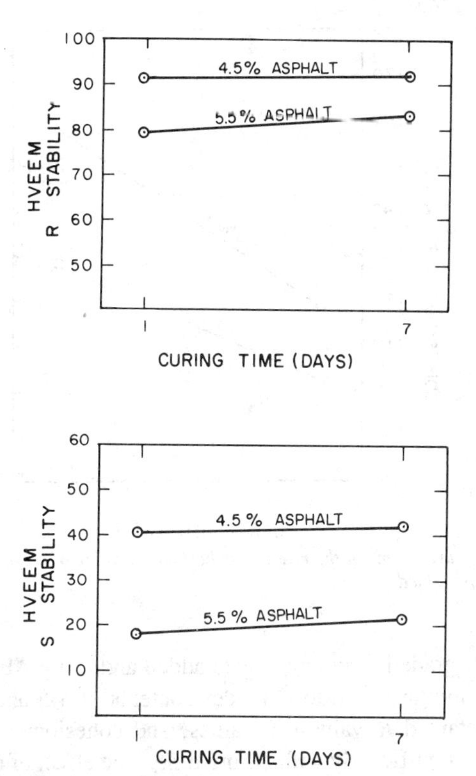

FIG. 7—*Effect of curing time on the retained Hveem* R *and* S *values of recycled mixtures with 25% well-graded aggregate added.*

Attenuation of Water Damage on Cold-Recycled Asphalt Mixtures

The results of this study indicate that the major factors affecting the water resistance of cold-recycled mixtures include (1) compactive effort, (2) curing time, (3) type of added virgin binder, and (4) type of added virgin aggregate. The proper consideration of these factors in the design of cold-recycled asphalt mixtures will help attenuate the water damages on these mixtures. The effects of these four factors on the water resistance of cold-recycled mixtures are presented here.

The effect of compactive effort on the R and S values of a recycled mixture under the dry and the wet conditions are shown in Fig. 6. The recycled mixture tested was made from old Pavement Material 2 using AE-150 as the added binder. It can be noted that the effect of water decreases as the compactive effort increases. It can be concluded that higher compactive efforts can be used to increase the water resistance of the cold-recycled asphalt mixtures.

The effect of curing time on the R and S values and on the cohesiometer values of two recycled mixtures under the wet condition are shown in Figs. 7 and 8, respectively. The two recycled mixtures were made from old Pavement Material 2,

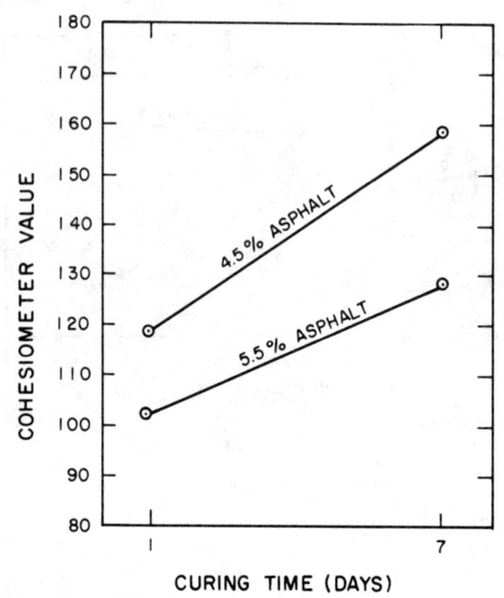

FIG. 8—*Effect of curing time of the retained cohesiometer value of recycled mixtures with 25% well-graded aggregate added.*

with 25% of well-graded virgin aggregate added and using AE-150 as the added binder. The two mixtures had total binder contents of 4.5 and 5.5%. It can be noted that the retained *R* values, *S* values, and cohesiometer values of these mixtures increase significantly with curing time. The effect of curing time on the retained resilient moduli of cold recycled mixtures have been shown previously in Fig. 3. There, it was noted that the retained resilient moduli at ultimate curing condition were significantly higher than those at one-day curing.

The effect of the type of added binder to the water resistance of a cold-recycled asphalt mixture is illustrated here by comparing a mixture using AE-150 as the added binder with a mixture using AE-90 as the added binder. Both mixtures were made from old Pavement Material 2 with 25% of well-graded virgin aggregate added and had a total binder content of 4.5%. The "wet" *S* values of these two mixtures are depicted in Fig. 9. It can be noted that at one-day curing the recycled mixtures using AE-150 as the added binder show higher retained *S* values than those of the mixtures using AE-90 as the added binder. However, the *S* values for the two mixtures converge after seven days. The effect of the type of added virgin binder to the water resistance of cold-recycled mixtures is significant and should be properly considered in the mix design.

The effect of the type of added virgin aggregate to the water resistance of a cold-recycled mixture is illustrated here by comparing a mixture using 25% coarse virgin aggregate with a similar mixture using 25% well-graded virgin aggregate. Both recycled mixtures were made from old Pavement Material 2 using AE-150 as the added binder. The retained *R* values of these two mixtures are shown in Fig. 10. It can be noted that the recycled mixtures with a coarse

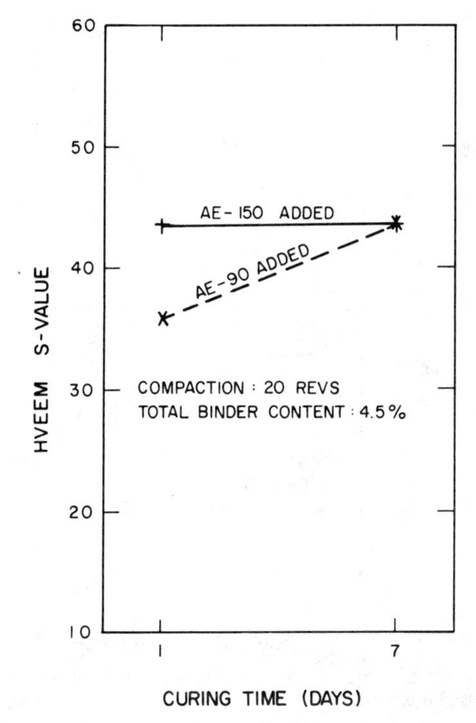

FIG. 9—*Effect of type of added virgin binder on the retained Hveem* S *value of recycled mixtures with 25% well-graded aggregate added.*

virgin aggregate have lower retained *R* values than the mixtures with a well-graded virgin aggregate. For this pavement material, adding a coarse virgin aggregate made the recycled mixtures harder to compact and subsequently less water-resistant than those obtained by adding a well-graded virgin aggregate. These results show the importance of the proper selection of added virgin aggregate for attenuation of water damage of cold-recycled asphalt mixtures.

Conclusions

Major conclusions from this study are summarized as follows:

1. The resilient modulus, Marshall stability, Hveem stabilometer, *S* value, and Hveem cohesiometer tests are effective in measuring the changes in stiffness, stability, and cohesion of cold-recycled mixtures caused by water damage.
2. The Hveem stabilometer *R* value test is not very sensitive in measuring the effect of water when the retained stability of a recycled mixture is relatively high. However, the *R* value test becomes very sensitive when a mixture is unstable, and thus it is very effective in determining the recycled mixtures that are highly susceptible to the action of water.

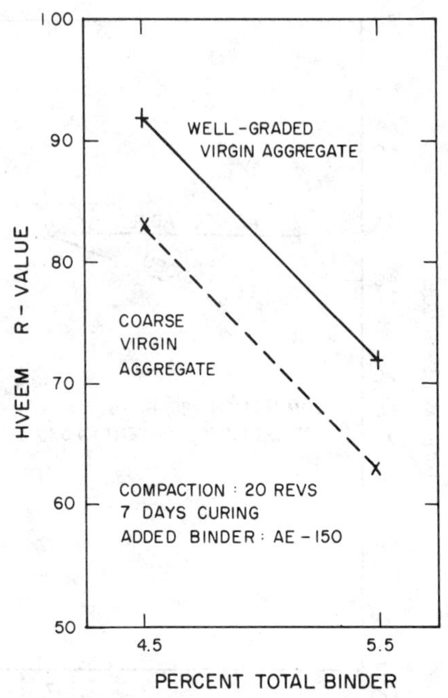

FIG. 10—*Effect of type of added virgin aggregate on the retained Hveem* R *value of recycled mixtures with 25% virgin aggregate.*

3. Water resistance of cold-recycled mixtures generally increases with curing time and compactive effort.

4. The effects of the type of added virgin binder and the type of added virgin aggregate to the water resistance of cold-recycled asphalt mixtures are significant and should be considered in the mix design.

Acknowledgments

The authors wish to express their appreciation to the Indiana Department of Highways and the Federal Highway Administration for their financial support of this study. The asphalt emulsions were supplied by K. E. McConnaughay, Inc. The asphalt cements were supplied by American Oil Company.

References

[1] Tia, M. "Characterization of Cold-Recycled Asphalt Mixtures," Report FHWA/IN/JHRP-82/5, Purdue University, West Lafayette, IN, Feb. 1982.

[2] Van Wijk, A. and Wood, L. E., "Use of Foamed Asphalt in Recycling of an Asphalt Pavement," Transportation Research Record 911, Washington, DC, 1983, pp. 96–103.

[3] Brennen, M., Tia, M., Altschaeffl. A., and Wood, L. E., "Laboratory Investigation of the Use of Foamed Asphalt for Recycled Bituminous Pavements," Transportation Research Record 911, Washington, DC, 1983, pp. 80–87.

[4] Gadallah, A. A., "A Study of the Design Parameters for Asphalt Emulsion Treated Mixtures," Report JHRP-76-30, Purdue University, West Lafayette, IN, Oct. 1976.

[5] Kennedy, T. W., Roberts, F. L., and Lee, K. W., "Evaluation of Moisture Effects on Asphalt Concrete Mixtures," Transportation Research Record 911, Washington, DC, 1983, pp. 134–143.
[6] Mamlouk, M. S., "Characterization of Cold Mixed Asphalt Emulsion Treated Bases," Report IN/JHRP-79-19, Purdue University, West Lafayette, IN, 1979.
[7] Tia, M., "A Laboratory Investigation of Cold-Mix Recycled Bituminous Pavements," Report JHRP-78-23, Purdue University, West Lafayette, IN, 1978.
[8] Kennedy, T. W., McGennis, R. B., and Roberts, F. L., "Investigation of Moisture Damage to Asphalt Concrete and the Effect on Field Performance–A Case Study," Transportation Research Record 911, Washington, DC, 1983, pp. 158–165.
[9] The Asphalt Institute, "Water Sensitivity Test for Compacted Bituminous Mixtures," The Asphalt Institute Laboratory, College Park, MD, 1975.
[10] Schmidt, R. J., "A Practical Method for Measuring the Resilient Modulus of Asphalt-Treated Mixes," Highway Research Board Record 404, Washington, DC, 1972, pp. 22–32.

Humberto Castedo,[1] *Christine C. Beaudoin,*[1] *Leonard E. Wood,*[2] *and A. G. Altschaeffl*[2]

A Laboratory Study of the Effectiveness of Various Admixtures on the Attenuation of Moisture Damage Upon Various Foamed Asphalt Mixtures

---

REFERENCE: Castedo, H., Beaudoin, C. C., Wood, L. E., and Altschaeffl, A. G., **"A Laboratory Study of the Effectiveness of Various Admixtures on the Attenuation of Moisture Damage Upon Various Foamed Asphalt Mixtures,"** *Evaluation and Prevention of Water Damage of Asphalt Pavement Materials, ASTM STP 899,* B. E. Ruth, Ed., American Society for Testing and Materials, Philadelphia, 1985, pp. 104–115.

ABSTRACT: Durability characteristics of certain foamed asphalt mixtures were established during this laboratory investigation. Durability was characterized by a water sensitivity test and cyclic freezing and thawing. The various foamed asphalt mixtures were evaluated for durability after the mixtures had been compacted into 10.16-cm (4.00-in.) diameter by approximately 6.35-cm (2.50-in.) high specimens, and cured.

Durability effects of different variables were determined in this laboratory study. These variables were foamed asphalt content (two levels for water sensitivity section, one level for freeze-thaw section), aggregate (three types, used in both sections), additives (three types, plus a set without additives), and additive content (two levels for each additive in the water sensitivity section, one level for each additive in the freeze-thaw section). One asphalt type, one mixing and testing temperature, one set of curing conditions, and one moisture content per aggregate were used.

Resilient modulus and modified Marshall stability tests were used to monitor durability characteristics of the mix in the water sensitivity section. Durability in the freezing and thawing section was monitored by pulse-velocity, resilient modulus, and modified Marshall stability tests. When lime was used as an additive, durability, strength, and longevity of the foamed asphalt mixtures were substantially improved. The improvement achieved with the addition of lime was such that a material generally less suitable for bituminous mix, such as outwash sand or pit-run gravel, may rival a material more suitable for bituminous pavement mix such as crushed limestone.

During this study there were similar rates of decline per freeze-thaw cycle for pulse-velocity and modified Marshall stability. Pulse-velocity, a nondestructive test, appears to be related to the destructive modified Marshall stability method. There seemed to be a good reproducibility of pulse-velocity values among similar specimens.

[1]Graduate instructors in research, School of Civil Engineering, Purdue University, West Lafayette, IN 47907.

[2]Professors, School of Civil Engineering, Purdue University, West Lafayette, IN 47907.

KEY WORDS: moisture, durability, asphalts, flexible pavements, bituminous mixtures, water sensitivity, additive, pulse-velocity, Marshall stability, Foamix

One of the most important components of today's transportation system is the highway. There is a continuing interest to improve these highways in order to make that part of the transportation system more cost effective. Traditional methods of improving highways and roads are undergoing changes to better meet constraints such as (1) limited amount of natural resources, (2) the level of acceptable pollution that is unavoidably created by the use of certain resources, and (3) man's limited budget, which must deal with the ever-increasing financial cost of certain resources.

A traditional method for improving highways is the stabilization of aggregates with bituminous materials that are largely obtained by the processing of crude petroleum. Petroleum is a limited resource, a fact which has become more apparent in recent years. Certain methods of paving with bituminous materials, such as cutbacks, cause higher levels of hydrocarbons to be released into the environment than other methods.

The cost of bituminous materials constantly increases because of the continuing scarcity of petroleum and tighter environmental controls. Other expenses include the heating of aggregates and bitumen for hot mix applications, the transportation of water for emulsified asphalt uses, and the amount of time (which is money) needed for aeration of emulsified and cutback applications.

The use of foamed asphalt as a binder material is once again receiving attention [*1–3*]. The process of generating foamed asphalt has been previously reported in the literature [*1–7*].

This study addresses the durability question. Two types of durability were investigated: (1) the effect of soaking in water and (2) the effect of cycles of soaking in water, freezing, and thawing, on the stability and strength of foamed asphalt specimens. Control specimens, that is, specimens that have not been subjected to soaking or freezing and thawing, were also made and monitored for stability and strength.

The stripping of asphalt from the aggregate because of the action of water is a common problem limiting the durability of asphalt concrete in general. Certain chemicals have been found to increase the adhesion of asphalt to aggregates [*8*]. Three additives were used in this study to determine if they have any effect in the durability of foamed asphalt specimens. Also, a series of specimens without any additives were used as a control set.

The additives used were calcium hydroxide (lime), silane, and Indulin®. Lime has been used successfully as an additive to asphalt concrete in the past [*8*]. In this study the lime was added in a slurry form to the finer aggregate particles. Silane, the name of a family of organic chemicals, has been recommended as a possible antistripping agent [*9*]. It can be either added directly to the asphalt or used as an aggregate pretreatment. In this study the silane was added to the water used to bring the aggregate to the specified mixing moisture content. Indulin, typical of several liquids that are marketed as antistripping agents, was added directly to the asphalt in the heating tank before it was foamed.

The effect of water and cyclic freeze-thaw on the strength and stability of the foamed asphalt mix specimens was monitored by means of several tests: resilient modulus, a modified Marshall stability test, and pulse-velocity measurements. These specimens were taken, in most cases, to disintegration. This was the point in which the surface of the specimen displayed evidence of many cracks and loose material so that it would not fit in the loading frame of the resilient modulus apparatus, or give correct pulse-velocity readings. The results of the "nondestructive" tests as determined from the resilient modulus and pulse-velocity test, were compared to the results of the modified, more common (but destructive) Marshall stability test. The purpose of this comparison was to establish the likelihood of a correlation between the modified Marshall method and a nondestructive test. Such a correlation might then be used as a gage of relative deterioration of a foamed asphalt specimen subjected to cycles of water soaking, freezing, and thawing.

It is hoped that the durability of foamed asphalt mixtures will be better understood, as a result of this study, so that this type of pavement mix can be seriously considered for field use.

Materials and Equipment

Aggregate Material

The aggregate materials used in this research study were outwash sand, pit-run gravel, and crushed limestone. These aggregates were utilized in previous foamed asphalt laboratory investigations [*1*,*4*,*5*]. Their gradations and other physical properties are presented in Tables 1 and 2 of this report.

Bituminous Material

A foamed AC-20 asphalt cement was used as the binder material for preparing the mixtures. The respective amounts used with each different aggregate were selected based on findings and results of previous studies conducted at Purdue University [*1*,*5*,*6*], CO [*4*], and data obtained by CONOCO Oil Co. [*7*], the developer of the laboratory foamed asphalt generator used in this study. The properties and foam characteristics of this binder material are listed in Table 3.

Additives

The three different types of additives used in this laboratory investigation were lime, silane, and Indulin. Additives were included in this work because of the history of poor performance by foamed asphalt mixtures with respect to moisture deterioration [*2*–*4*].

TABLE 1—*Gradation of aggregates, % passing (from dry sieve analysis).*

Sieve Size	Pit-Run Gravel	Outwash Sand	Crushed Stone
1 in. (25.0 mm)	100.0	100.0	100.0
¾ in. (19.0 mm)	90.0	100.0	96.7
½ in. (12.5 mm)	84.0	99.7	79.0
⅜ in. (9.5 mm)	76.0	99.5	66.0
No. 4 (4.75 mm)	65.0	97.6	43.5
No. 8 (2.36 mm)	50.0	94.3	34.0
No. 16 (1.18 mm)	35.0	90.7	25.0
No. 30 (600 μm)	20.0	78.4	17.1
No. 50 (300 μm)	8.0	38.3	14.0
No. 100 (150 μm)	5.0	9.5	11.5
No. 200 (75 μm)	3.0	2.4	9.0

TABLE 2—*Properties of aggregates.*

Properties	Bulk Sp Gr (SSD)[a]	Apparent Sp Gr[a]	Absorption, %[a]	Mineral Filler[b]
Pit-run gravel	2.644	2.710	1.56	nonplastic
Outwash sand	2.607	2.707	1.20	nonplastic
Crushed limestone	2.696	2.741	1.28	nonplastic

[a]ASTM Test Method for Specific Gravity and Absorption of Coarse Aggregate (C 127).
[b]AASHTO Test Method for Determining the Plastic Limit and Plasticity Index of Soils (T90-70).

TABLE 3—*Properties of asphalt cement AC-20.*

Property	Value
Penetration (0.1 mm), 100 g, 5 s, 25°C	42
Softening point, ring and ball	48°C (118°F)
Ductility, 25°C (77°F), 5 cm/min	150 cm+
Kinematic viscosity, 150°C (302°F)	229 cSt
Kinematic viscosity, 160°C (320°F)	126 cSt
Kinematic viscosity, 180°C (355°F)	72 cSt
Foam characteristics	
expansion ratio	13
half life	14 s

The most common way of handling the durability problem of foamed asphalt mixtures in the field is by assuring that the foamed asphalt layer is drained and sealed with a hot asphalt surface mix layer [*1*,*2*]. However, if an effective and economical additive could be found, which will improve foamed asphalt mixtures' resistance to deterioration, then foamed asphalt would be a more viable alternative to other forms of bitumen stabilization.

A detailed listing of the additives' characteristics and properties is presented in Tables 4, 5, and 6.

TABLE 4—*Silane properties* [13].

Property	Value
Appearance	clear liquid (CTM 0176)[a]
Color	light straw to yellow (CTM 0176)
Viscosity	6 cSt (CTM 0004)
Specific gravity (25°C)	1.02 g/mL (CTM 0001A)
Flash point, Cleveland open cup	127°C (260°F) (CTM 0006)

[a]CTM stands for Corporate Test Method, which generally correspond to ASTM standard test methods. CTMs are available from Dow Corning upon request.

TABLE 5—*Calcium hydroxide analysis.*

Parameters	% by Weight
Calcium hydroxide	
$Ca(OH)_2$	FW = 74.09[a]
Chloride (Cl)	0.01
Iron (Fe)	0.05
Sulfur compounds	0.06
Heavy metals (as lead)	0.002
Magnesium and alkali salts	0.62
Insoluble in dilute hydrochloric acid	0.006

[a]FW is formula weight.

TABLE 6—*Indulin properties* [14].

Property	Value
Active ingredients	100%
Form	viscous liquid
Color	dark brown
Specific gravity, 25 C/50 C	0.970/0.963
Pour point	13°C (55°F)
Viscosity, SSF 50°C (122°F)	80
Viscosity, CPS 25°C (77°F)	2450
Viscosity, CPS 50°C (122°F)	150
Flash point, Cleveland open cup	160°C (320°F)
Fire point, Cleveland open cup	168°C (335°F)
Weight/gal 25°C (77°F)	3.67 kg (8.10 lb)

Equipment

A list of the laboratory equipment used in this investigation is as follows:

- Foamed Asphalt Generator [*6*,*7*]
- California Kneading Compactor [*10*]
- Compression Machine, Mechanical Mixer, Ovens, and Scales [*6*]
- Vacuum Saturation Apparatus [*1*]
- Freeze-Thaw Apparatus [*1*]

- Pulse-Velocity Apparatus [*11*]
- Diametral Resilient Modulus Device [*6*]
- Marshall Testing Apparatus [*10*]

More information on the various pieces of equipment used to prepare and test the foamed asphalt specimens can be found in the respective references listed in brackets next to each item.

Experimental Study

The determination of the durability characteristics of foamed asphalt mix specimens was the main purpose of this laboratory study. These characteristics were investigated by conducting a water sensitivity test and a freeze-thaw test. Untested specimens were used as a control set.

Water Sensitivity Test

The water sensitivity test has been used before on foamed asphalt mix specimens, but the procedure used was rather severe [*6*]. A less harsh test recommended by Ruckel et al [*3*] was used. The procedure is similar to the one found in the *Asphalt Emulsion Manual* (MS-19) of the Asphalt Institute [*10*].

Freeze-Thaw Test

Resilient modulus and pulse-velocity results were obtained from test specimens before performing each vacuum saturation and freezing-thawing cycle. After the last freeze-thaw cycle was completed, the specimens were tested in the Marshall stability apparatus.

Control Testing

Each combination of the variables involved in this study, required the formation of six specimens to be used in each of the durability tests (water saturation, and water saturation plus freeze-thaw tests). Three of these six specimens were used as control specimens in both durability sections. In other words, immediately after curing, these specimens were subjected to the full battery of testing and thus were destroyed (by the modified Marshall stability test) before any vacuum saturation, freezing, or thawing could occur. The results of a given set of three specimens were averaged together to yield the "original" values of those mixtures.

Testing Routine

The effects of the durability tests and the additives on the foamed asphalt mix specimens were determined by the following procedures. The foamed asphalt mix specimens were prepared, cured (24 h in mold, and 24-h oven at 40°C [104°F]), weighed, and measured for height before pulse-velocity testing. (The pulse-velocity test was used only in the freeze-thaw study). The control-test specimens were measured for pulse-velocity once before being subjected to the modified Marshall stability test.

The pulse-velocity procedure used in this laboratory study was performed following ASTM Test Method for Laboratory Determination of Pulse Velocities and Ultrasonic Elastic Constants of Rock (D 2845). More details on this non-destructive test can be found in Ref *11* of this report.

The water sensitivity and the freeze-thaw studies made use of the resilient modulus test. The diametral resilient modulus test was similar to the procedure developed by Schmidt with some modifications [*1*,*6*]. A pulsating load of 220 N (50 lb) was applied across the vertical diameter of the specimen with a dwell time of 0.1 s, every 3 s. Vertical deformations of the specimen were plotted on a strip chart recorder from which the instantaneous resilient modulus of the specimen was obtained.

The modified Marshall stability test was used in both the water sensitivity and freeze-thaw studies. The foamed asphalt mix specimens were prepared, cured, and measured for height and weight before modified Marshall stability testing. The term "modified" is used to indicate a test temperature of approximately 22°C (72°F), which is different than the standard test temperature. The rest of the Marshall test was performed according to ASTM Test Method for Resistance to Plastic Flow of Bituminous Mixtures Using Marshall Apparatus (D 1559).

Test Results

Water Sensitivity Test

The modified Marshall stability results for pit-run gravel and outwash sand foamed asphalt specimens displayed definite trends in the water sensitivity section. It was more difficult to determine trends for the crushed stone specimens. Crushed limestone is an inherently strong material, and additive effects in this type of foamed asphalt mixtures were minimal.

Significantly higher percentages of retained modified Marshall stabilities were observed in specimens treated with additives after undergoing the water sensitivity test. The silane and Indulin additives appeared to have had a weakening effect on the control specimens (those specimens not soaked in water). After the water sensitivity test, however, the silane and Indulin additives cause these specimens to retain higher modified Marshall stabilities than the specimens without additives. Specimens with lime added benefited both before and after the water sensitivity test. Specimens with lime were as strong or stronger than specimens without additives before the water sensitivity test.

Figure 1 depicts the modified Marshall stability values for pit-run gravel specimens with 4.00% foamed asphalt content, both before and after the water sensitivity test. This graph is representative of the trends observed with the rest of the specimens.

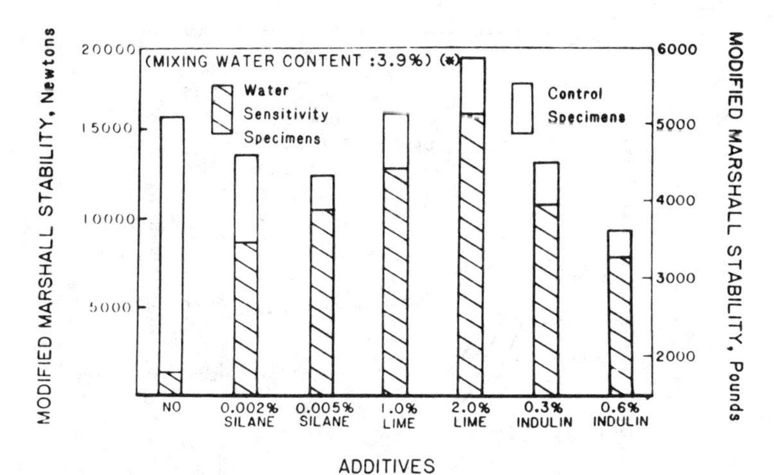

FIG. 1—*Modified Marshall stability for pit-run gravel mixtures with 4% (*) foamed asphalt content. (* is percent by dry weight of aggregate).*

Past experience shows that lime has performed favorably as an additive to asphalt cement mixes [*12*]. This good performance is more likely for the following two reasons: (1) lime is a good antistripping agent. It tends to make the aggregate more hydrophobic and thus the asphalt can adhere more readily to the aggregate even in the presence of water. (2) When used in small quantities, such as in this study, lime, being a fine-grained material, serves to stiffen the asphalt matrix so that the stability increases without the specimen becoming brittle.

Silane and Indulin both increase a foamed asphalt mix specimen's resistance to water by improving the bond between the asphalt and the aggregate. There were two levels used for each of the additives in the water sensitivity section. The modified Marshall stabilities for one type of additive were generally fairly close in value at the two levels for that type of additive.

The higher level of Silane (0.005 versus 0.002%) usually gave a higher stability than the lower level specimens. For the lime specimens, the higher level (2.0 versus 1.0%) usually had a slightly higher stability than the lower level. For the Indulin specimens, the lower level (0.3 versus 0.6%) usually had a slightly higher stability than the higher level of this additive. A more complete battery of tests has to be conducted to determine the optimum level of each additive to be used with the different aggregates.

The resilient modulus test did not indicate any clear trend in the water sensitivity section of this study.

Freezing and Thawing Test

Comparable trends were observed in the freeze-thaw section for modified Marshall stability and pulse-velocity values. Crushed stone specimens were again

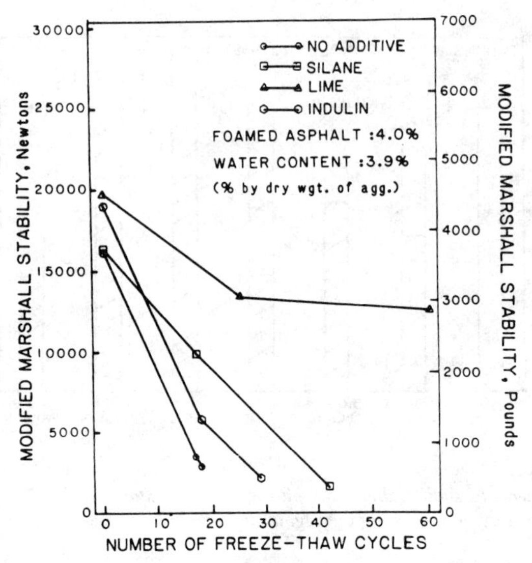

FIG. 2—*Modified Marshall stability versus number of freeze-thaw cycles for pit-run gravel mixtures.*

strong so that the effect of the additives was minimal. Trends and effects were much more visible with the pit-run gravel and outwash sand mixtures.

Specimens without additives usually had the lowest stability and pulse-velocity values, and disintegrated the fastest. Specimens with additives had increased longevity.

Pit-run gravel and outwash sand specimens, treated with lime, silane, and Indulin, had markedly improved modified Marshall stabilities over the non-additive specimens. However, lime was by far the best of the three additives used in this part of the study. Graphs of modified Marshall stabilities versus number of freeze-thaw cycles for all three aggregates are presented in Figs. 2, 3, and 4 (pit-run gravel, outwash sand, and crushed stone, respectively).

The lime-treated specimens gave consistently high pulse-velocity values. Pit-run gravel and outwash sand specimens with lime had by far the highest number of freezing and thawing cycles to disintegration (refer to the Introduction). In fact, the pit-run gravel and outwash sand specimens treated with lime were destroyed by the Marshall stability test prematurely in the interest of time. They could have easily existed far beyond the 60 cycles indicated in Fig. 5 where number of freeze-thaw cycles are plotted against type of additive. The success of lime as an additive for foamed asphalt mixes can be attributed to the characteristics of lime mentioned before. These beneficial characteristics were also observed in the freeze-thaw section of this laboratory study.

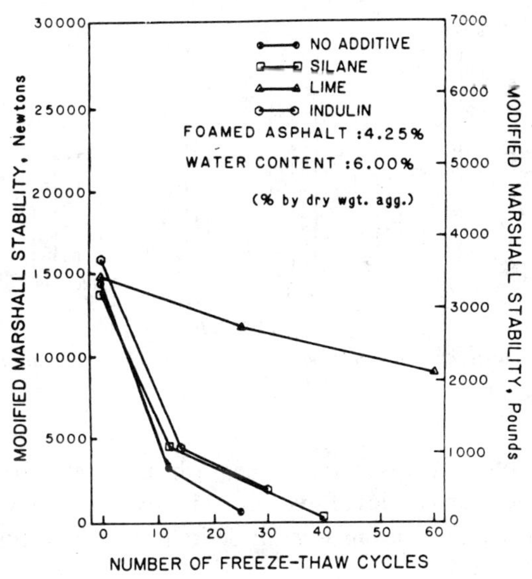

FIG. 3—*Modified Marshall stability versus number of freeze-thaw cycles for outwash sand mixtures.*

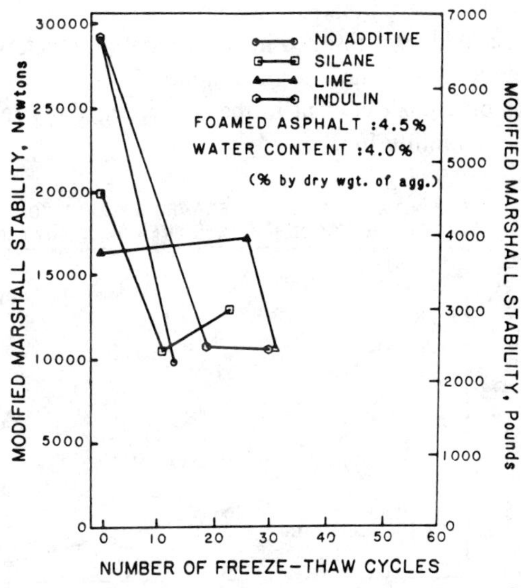

FIG. 4—*Modified Marshall stability versus number of freeze-thaw cycles for crushed stone mixtures.*

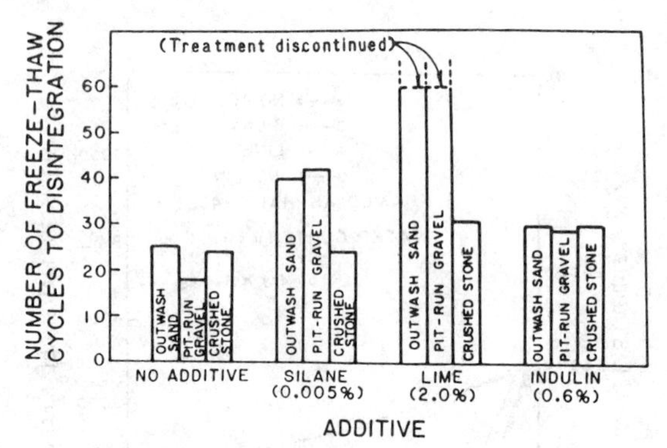

FIG. 5—*Longevity graph: number of freeze-thaw cycles before disintegration.*

Conclusions

The durability of foamed asphalt specimens when subjected to a modified form of vacuum saturation and cycles of freezing and thawing were investigated in this study. The following conclusions were made from an examination of the results described in this report:

- The vacuum saturation test weakened the foamed asphalt specimens.
- Additives had little effect on the crushed stone specimens in the water sensitivity test. However, the additives enabled outwash sand and pit-run gravel foamed asphalt samples to retain much more of their original stability after vacuum saturation than the untreated specimens.
- Lime and silane additives resulted in better durability than would have been obtained with the addition of more asphalt alone.
- Higher levels of asphalt content in the water sensitivity test generally resulted in increased durability [*6*].

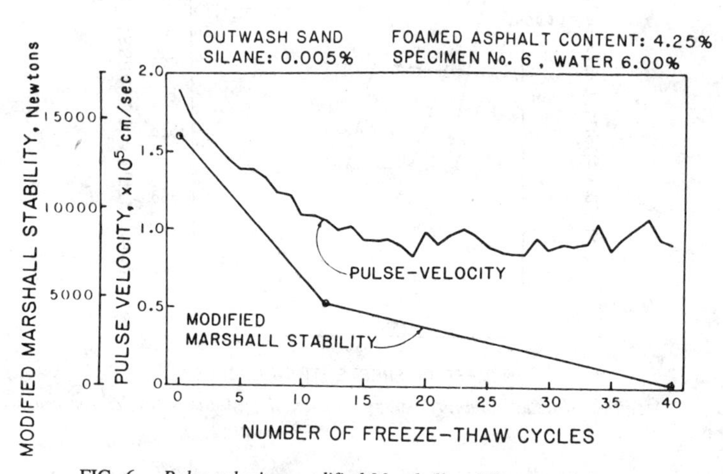

FIG. 6—*Pulse-velocity: modified Marshall stability comparison.*

- Cyclic vacuum saturation, freezing, and thawing weakened the foamed asphalt specimens.
- The best performing mixtures were obtained using lime as an additive. The improvement in the Marshall stability values, stability retention, and specimen longevity is such that a material generally less suitable for bituminous mix, such as outwash sand or pit-run gravel, gains so much from the addition of lime as to rival a better material such as crushed limestone. The outwash sand and pit-run gravel specimens had somewhat lower stabilities than the lime-treated crushed stone specimens, but they were able to withstand many more cycles of freezing and thawing before disintegrating. Silane and Indulin appear to yield favorable results in terms of stability retention; however, the results were not so marked when compared with the values obtained for the lime-treated specimens.
- According to the results obtained in this study, pulse-velocity is a nondestructive test that appears to be related with the modified Marshall stability (Fig. 6). The similar rate of decline observed between these two tests also shows that there is good reproducibility of results between similar specimens.

References

[1] Wood, L. E., Altschaeffl, A. G., Cravens, C. M., and Castedo, H., "The Use of Foamed Asphalt in Bituminous Stabilization of Base and Subbase Materials and Recycled Pavement Layers," Joint Highway Research Project, Purdue University, West Lafayette, IN, Final Report JHRP-84-5, March 1984, 200 pp.

[2] Lee, D. Y., "Treating Marginal Aggregates and Soil with Foamed Asphalt," Engineering Research Institute, Iowa State University, Ames, IA, 1980, 32 pp.

[3] Ruckel, P. G., Acott, S. M., and Bowering, R. H., "Foamed Asphalt Paving Mixtures: Preparation of Design Mixes and Treatment of Test Specimens," Presentation, Session 174, Annual TRB Meeting, Washington, DC, Jan. 1982.

[4] Abel, F., "Foamed Asphalt Base Stabilization," 6th Annual Asphalt Paving Seminar, Colorado State University, Fort Collins, CO, Dec. 1978.

[5] Brennan, M., Tia, M., Altschaeffl, A. G., and Wood, L. E., "A Laboratory Investigation on the Use of Foamed Asphalt for Recycled Bituminous Pavements," Report JHRP-81-5, Joint Highway Research Project, Purdue University, West Lafayette, IN, March 1981, 48 pp.

[6] Castedo Franco, L. H., "The Use of Foamed Asphalt in Bituminous Stabilization of Base and Subbase Materials," Master's thesis, Purdue University, West Lafayette, IN, Dec. 1981.

[7] Continental Oil Company (CONOCO), "Foamix-Foamed Asphalt, An Alternative Mixing Method," Technical Report, 1979.

[8] Oglesby, C. H. and Hicks, R. G., *Highway Engineering,* 4th ed., John Wiley and Sons, New York, 1982, 844 pp.

[9] DiVitto, J. A. and Morris, G. R., "Silane Pretreatment of Mineral Aggregate to Prevent Stripping in Flexible Pavements," Arizona Transportation Research Center, Arizona State University, Tempe, AZ, 1982.

[10] The Asphalt Institute, "A Basic Asphalt Emulsion Manual," College Park, MD, Manual Series 19, March 1979.

[11] Manke, P. G. and Gallaway, B. M., "Pulse Velocities in Flexible Pavement Construction Materials," Highway Research Record 131, Washington, DC, 1966, pp. 128–153.

[12] Davison, D. T. and Handy, R. L., "Lime and Lime-Pozzolan Stabilization," Highway Engineering Handbook, Part 4, Section 21, K. B. Woods, Ed., McGraw-Hill, New York, 1960, pp. 21-98–21-100.

[13] Dow Corning Corporation, "New Product Information—Dow Corning 990 Asphalt Additive," Technical Report, Dow Corning Corporation, Midland, MI, 3 pp.

[14] Westvaco Technical Bulletin, "Indulin AS-1," Technical Report 32078, Westvaco Chemical Division, Polychemicals Department, Charleston Heights, SC.

Recommendations for the Evaluation and Use of Lime Additives

Thomas W. Kennedy [1]

Prevention of Water Damage in Asphalt Mixtures

REFERENCE: Kennedy, T. W., **"Prevention of Water Damage in Asphalt Mixtures,"** *Evaluation and Prevention of Water Damage to Asphalt Pavement Materials, ASTM STP 899,* B. E. Ruth, Ed., American Society for Testing and Materials, Philadelphia, 1985, pp. 119–133.

ABSTRACT: This paper summarizes the findings of a six-year study of moisture damage of asphalt mixtures, which was conducted at The University of Texas at Austin. The objectives of the study were to define the nature and severity of moisture damage in asphalt pavement mixtures, develop techniques to identify mixtures that are moisture susceptible, and develop recommendations to eliminate or minimize the problem. While this study has primarily focused on problems and mixtures in Texas, additional mixtures and experience from other areas in North America were considered. The study involved laboratory investigations, including the evaluation of mixtures subsequently used in construction, a field evaluation of methods of treating asphalt mixtures with hydrated lime, and an evaluation of actual pavement mixtures that exhibited moisture damage.

The majority of the work utilized the wet-dry indirect tensile test (Lottman), the Texas freeze-thaw pedestal test, and the Texas boiling test, which are discussed with respect to their ability to estimate moisture susceptibility of asphalt-aggregate mixtures. The results obtained for a variety of antistripping agents, including hydrated lime and silanes, are summarized, along with the results of a field experiment to evaluate methods for using hydrated lime. Finally, recommendations are made with regard to methods of alleviating moisture damage, including methods of treating moisture susceptible aggregate with hydrated lime.

KEY WORDS: asphalts, aggregates, stripping (distillation), calcium oxides, tests, asphalt mixtures, moisture damage, antistripping additives

During the past few years, asphalt pavement mixtures have suffered extreme damage because of the adverse effects of moisture. In recognition of these problems a research study, Project 3-9-79-253, "Moisture Effects on Asphalt Mixtures," was initiated at The University of Texas at Austin. This paper summarizes the highlights of the overall study, which is detailed in eight research reports [*1–8*].

[1]Associate dean of engineering for research and planning and Engineering Foundation professor, Office Dean of Engineering, University of Texas at Austin, Austin, TX 78712-1080.

Moisture Damage

Moisture damage occurs in two forms, softening and stripping. Softening is characterized by a reduction of cohesion, which produces a reduction in strength and stiffness of the asphalt mixture. Stripping, on the other hand, involves a loss of adhesion and the physical separation of the asphalt cement and aggregate primarily caused by the action of moisture [*1*]. A similar separation can also occur because of surface coatings on the aggregate or to smooth aggregates with minimal surface texture.

Preliminary evidence of stripping of asphalt pavement mixtures often occurs as localized instability and patch flushing or bleeding, that is, localized shiny areas. Flushing occurs when a portion of the stripped asphalt cement rises to the surface of the pavement, producing localized shiny areas of asphalt. This bleeding is not necessarily confined to the wheel paths but rather is often distributed across the pavement surface. Deformations in the form of shoving and rutting may also develop because of the loss of structural strength and stiffness and because of instability caused by the excessive amounts of asphalt near the surface.

It also may be found that cores cannot be obtained because of the lack of cohesion and strength in the lower portion of the pavement layers. During examination of the asphalt-aggregate mixture it may appear that the aggregates are completely clean, without evidence of asphalt; however, small beads of asphalt may be evident, and reheating and mixing may cause recoating of the aggregate. Thus, most of the asphalt is apparently still physically present.

The basic cause of stripping must be identified in order to select the method of treatment. Unfortunately, no single mechanism of stripping has been universally accepted, and it is possible that different mechanisms occur for different conditions and that more than one mechanism may actually produce failure. Because of the number of proposed mechanisms and the lack of agreement, the causes of stripping were categorized as physical-chemical reactions, surface coatings, and smooth surface textures [*1*].

Surface coatings on the aggregate prevent adequate adhesion with the asphalt cement and can be reduced by washing the aggregate before use. Smooth aggregates also minimize the ability of the aggregate and asphalt to develop adequate adhesion. Crushing of the aggregate, which tends to produce surfaces with more texture, may therefore also reduce stripping. Physical-chemical reactions, however, are of primary concern and probably will require other treatments to alleviate stripping.

Moisture Damage in Texas

A survey of Texas [*1*] and the experiences in other states indicated that the extent and severity of moisture damage of asphalt mixtures is primarily related to the environment, aggregate, asphalt, and mixture properties.

As expected, areas with high rainfall and high water tables experienced a much greater amount of moisture damage and stripping. Most of the moisture problems in Texas occurred in the eastern and southeastern part of the state, which has a high water table and high rainfall, approaching 1270 mm/year (50 in./year). However, isolated cases occurred in west Texas and were generally correlated with the high negative values of the Thornthwaite moisture index [*1*], which indicate a high potential for attracting moisture [*9*].

Siliceous river aggregates and rhyolite were found to have a greater propensity for stripping than other Texas aggregates. Therefore, areas that use large quantities of siliceous river and rhyolite aggregates and have relatively high amounts of rainfall experience greater moisture-related distress problems. While Texas limestones are generally resistant to stripping, experience indicates that some limestone type aggregates are susceptible to moisture damage.

The type and viscosity of asphalt also are important. Certain asphalts used in Texas produce mixtures with a greater resistance to stripping. In addition, it has been shown that higher viscosity asphalts are more resistant to stripping.

Methods of Treatment

The following procedures, treatments, and methods of protection can reduce the moisture damage and related pavement distress [*3,10,11*]:

(1) utilize a properly designed mixture,
(2) provide adequate compaction,
(3) eliminate the use of moisture-susceptible aggregates and asphalts,
(4) provide adequate drainage,
(5) seal the asphalt-aggregate mixture surfaces, and
(6) treat the moisture-susceptible aggregates and asphalt.

The amount and ease with which moisture can enter an asphalt concrete mixture is dependent on the asphalt content and aggregate gradation. Dense, well-graded mixtures with the proper asphalt content will more effectively prevent moisture penetration.

Adequate compaction will reduce the air voids and the continuity of the air-void system. This prevents the penetration of moisture into the mixture, thus reducing the possibility for stripping to occur. The air-void content should, ideally, be less than 7%. At void contents in excess of 7%, water can readily penetrate the mixture. Thus, compaction should achieve a relative density of at least 93% of the theoretical maximum density.

It may be desirable to eliminate the use of certain moisture-susceptible aggregates and, to a lesser extent, certain asphalts. Such an approach may be costly, especially in areas with limited aggregate and asphalt sources. Nevertheless, in view of the long-term maintenance requirements, reduced pavement life and performance, and in some cases, the rapid and severe failure of the pavement, it may in reality be the most economical solution if adequate protection cannot be achieved or if the mixture cannot be adequately protected.

Drainage should be provided to eliminate moisture, which causes stripping to occur. This involves rapid removal of surface water and prevention of moisture movement into the mixture from the subgrade, subbase, and base by drainage and by maintaining a sufficient pavement elevation above the water table. The use of open-graded friction courses has been shown to cause stripping by allowing moisture to enter the underlying layers under the action of traffic, especially if the moisture cannot readily drain laterally.

Both the top and the bottom surface of the asphalt mixture can be sealed to prevent moisture penetration. This approach requires that careful consideration be given to the source of moisture to avoid the possibility of trapping water in the mixture. A number of cases in Texas and other states have been reported in which a surface seal was placed on an existing roadway resulting in subsequent rutting and deterioration caused by stripping. Thus, sealing of the bottom surface may trap surface water by preventing drainage into the underlying layers, and similarly, surface sealing may prevent evaporation of moisture from underlying layers, which is moving upward through the mixture.

The aggregates and asphalts for mixtures that are susceptible to stripping can be treated with a number of antistripping additives. These additives commonly include the following:

Liquid Antistripping Agents

These materials, which are commercially available under various trade names and designations, are normally added to the asphalt cement and have been used for a number of years.

Portland Cement

Portland cement is added to the aggregate and has been reported to be generally effective; however, except for a limited number of states, it has not been used widely.

Hydrated Lime

Hydrated lime normally is added to the aggregate and has been used widely in portions of the United States during various times in the past.

Test Methods

At the beginning of the study, numerous tests and test variations had been proposed and were being used to evaluate the moisture susceptibility of asphalt-aggregate mixtures, with and without additives. It appeared, however, that many of the tests being used were not satisfactory and that new methods were needed. After review and evaluation of these various tests, the indirect tensile test with moisture conditioning [*7*], Texas freeze-thaw pedestal test [*6*, *10*, *12*], and Texas boiling test [*4*, *10*, *13*] were adopted for use in the project.

Indirect Tensile Test on Dry and Wet Specimens

The indirect tensile test subjects a cylindrical specimen to compressive loads distributed along two opposite generators that create a relatively uniform tensile stress perpendicular to and along the diametrical plane, which contains the applied load and causes a splitting failure [*14*]. For proper evaluation, mixtures should contain about 7% air voids, and it has been recommended that compacted specimens should be conditioned to produce a relatively uniform degree of saturation in the range of 55 to 80% [*15*]; Texas currently recommends a range of 60 to 80% [*16*]. Moisture susceptibility is determined by the ratio of tensile strength in a wet condition to the tensile strength in a dry condition, that is, the tensile strength ratio. A recommended test procedure based on this study is contained in Ref *7*.

Texas Freeze-Thaw Pedestal Test

The pedestal test, which is a modification of a water susceptibility test originally developed at the Laramie Energy Technology Center [*17*], involves subjecting miniature asphalt-aggregate briquets to repeated freeze-thaw cycles (15 h at −1.2°C [10°F] and 9 h at 49°C [120°F]) while submerged in distilled water. The briquets, which contain a uniform aggregate size, are highly permeable, allow easy penetration of water, and minimize mechanical interlocking of the aggregate particles. Thus, the briquet properties are largely determined by the asphalt-aggregate bond and, to a lesser extent, the cohesion provided by the asphalt. Moisture susceptibility of an asphalt concrete mixture is evaluated by determining the freeze-thaw cycles required to crack a briquet seated on a beveled pedestal. Details of the test procedure are described in Refs *6* and *10*.

Texas Boiling Test

In this test, which is based on a review and evaluation of boiling tests that have been performed by various agencies [*18–20*], a visual observation is made of the extent of stripping of the asphalt from aggregate surfaces after the mixture has been subjected to the boiling action of water for a specified time. The extent of stripping is rated visually and compared to a standard set of mixtures, which vary from 0 to 100% of the asphalt cement retained. Based on field performance, mixtures that retain less than 70% of the asphalt cement are considered to be moisture susceptible. Details of the test procedure are described in Refs *4* and *10*.

Application of Tests

The wet-dry indirect tension test provides an evaluation of the mixture with the proper proportions of aggregates and asphalt and at a density configuration simulating field asphalt aggregate mixtures. The test is relatively easy and requires approximately two days to conduct; however, the Texas method, TEX 531-C [*16*], requires five days. The results are sensitive to differences in moisture content, and additional work may be necessary to define the required range of saturation. The Texas pedestal test can be used to evaluate the moisture susceptibility of the mineralogy of the combined aggregates and the asphalt or the individual aggregate components and the asphalt; however, it cannot evaluate the combined or individual gradations of the aggregate. Experience in Texas would indicate that the test is relatively accurate; however, the testing procedure is time consuming and thus is more applicable to preconstruction evaluations. In addition, it is difficult to determine the amount of antistripping additive required since the amount depends on surface area. The Texas boiling test, while probably not as accurate, is a very quick test to conduct and thus can be used easily during construction. In addition, the combined aggregates and asphalt or the individual aggregate components and asphalt can be evaluated.

Experience indicates that while there is a general correlation between test results obtained using the three tests, differences do occur that in some cases can be significant. Generally it has been found that liquid additives perform better with the boiling test while hydrated lime performs better with the pedestal and indirect tensile tests. Thus, possibly more than one test should be used.

Testing should be conducted on mixtures containing the aggregates and, to the extent possible, the asphalt cement to be used. The indirect tensile test and the boiling test can and should be conducted on the mixtures produced during construction.

Evaluation of Antistripping Additives

At the beginning of the study, various antistripping additives were being used in many states; however, moisture damage in the form of stripping was still occurring. Thus, a study of various liquid antistripping additives and hydrated lime was undertaken.

The early evaluations consisted of a formalized test program and a number of small case study experiments conducted on selected aggregate-asphalt combinations that were scheduled for use on actual highway construction projects. The antistripping agents evaluated were selected by district highway personnel. The results of the formal test program are contained in Refs *2* and *12*.

Only silanes and lime slurry were effective in improving moisture resistance as measured by the pedestal tests. Similar results were obtained with the boiling test. However, test results for individual project evaluations [*8*] involving a range of aggregates suggested that certain liquid additives were effective with specific combinations of asphalt and aggregate.

One very important finding was that each combination of asphalt, aggregate, and antistripping additive must be evaluated to determine whether the combination is resistant to stripping and that certain asphalts are more resistant to stripping than others. Thus, while stripping is primarily aggregate related, it is also dependent on the asphalt and the interaction of the asphalt and aggregate. Equally important is the fact that the effectiveness of antistripping additives is dependent on the specific combination of aggregate and asphalt (Fig. 1).

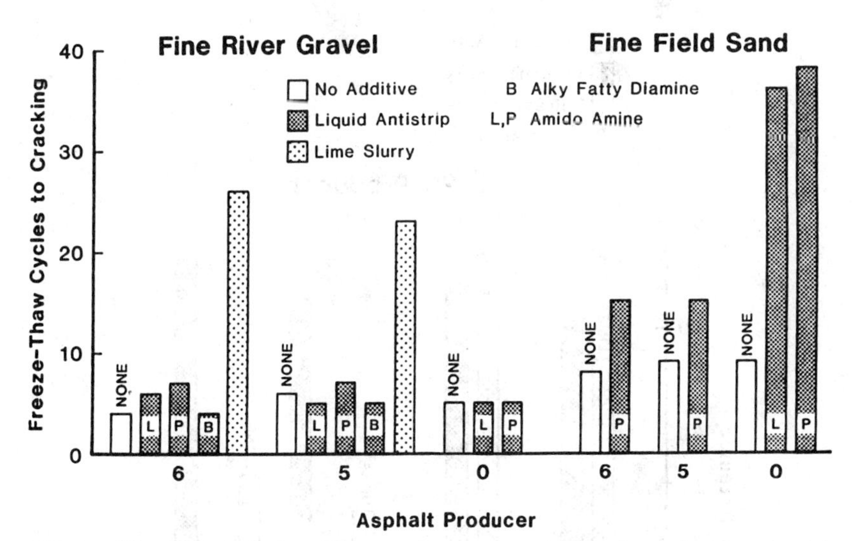

FIG. 1—*Effects of antistripping additives on the stripping resistance of asphalt mixtures involving two aggregates and three asphalt producers.*

As hydrated lime began to be used or considered as an antistripping additive, there was renewed interest in developing new or improved liquid antistripping additives. Thus, a limited test program was conducted and a number of these additives, which were proposed for use on actual projects, were evaluated. Typical results of the test program are summarized in Figs. 2 and 3. As shown,

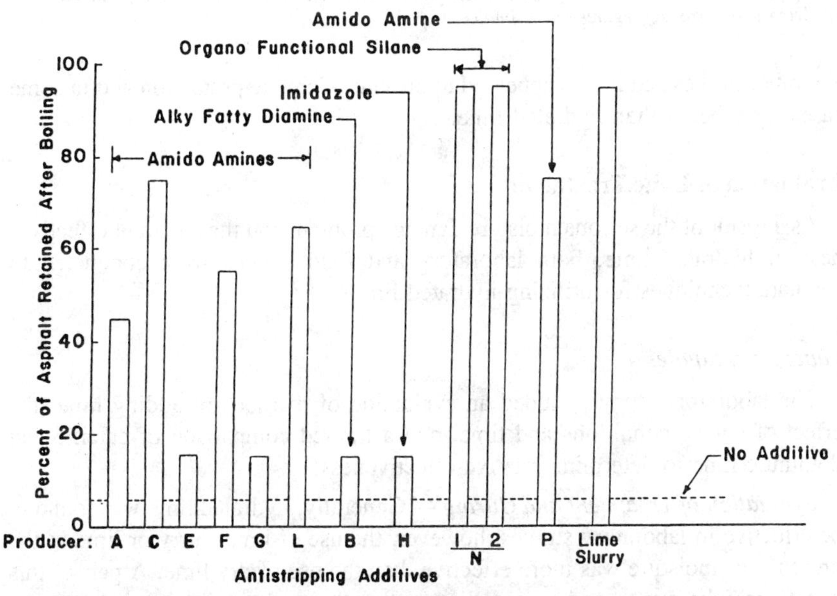

FIG. 2—*Effect of antistripping additives on the moisture resistance of asphalt mixtures.*

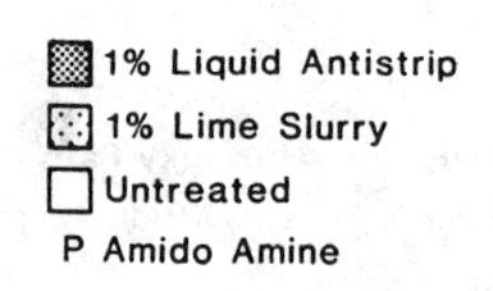

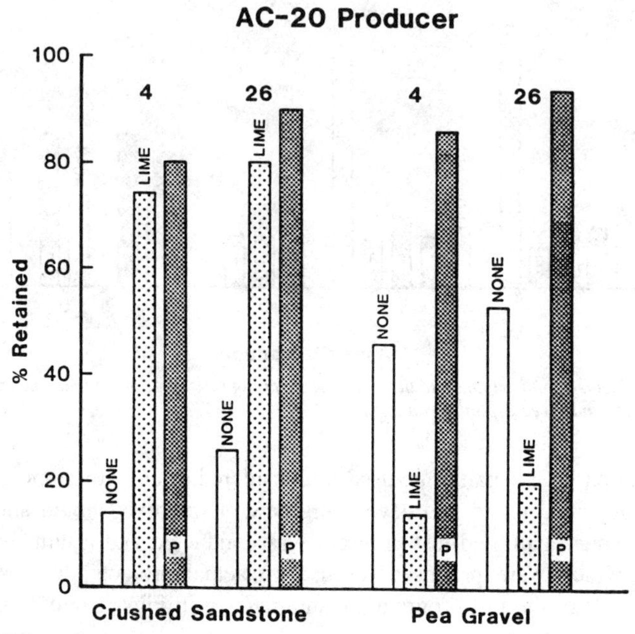

FIG. 3—*Effect of new antistripping additive on the moisture resistance of asphalt mixtures involving two Texas aggregates and asphalts.*

a number of these additives appeared to provide adequate protection and in some cases were better than hydrated lime.

Evaluation of Lime Treatment

As a result of the serious moisture damage problem and the apparent effectiveness of hydrated lime, both laboratory and field studies were conducted to evaluate techniques for utilizing hydrated lime.

Laboratory Studies

The laboratory study included an evaluation of methods of adding lime, the effect of curing conditions and time, and a limited comparison of calcitic and dolomitic lime to determine relative effectiveness.

Evaluation of Treatment and Curing—Generally, hydrated lime was found to be effective in laboratory studies; however, the use of lime slurry or lime in the presence of moisture was more effective than the use of dry lime. A part of this benefit may be due to an improved interaction between the lime and aggregate;

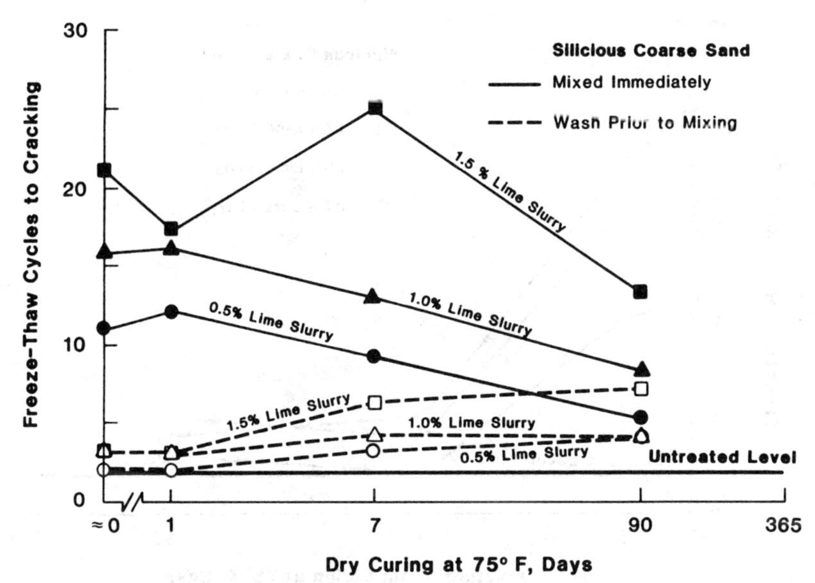

FIG. 4—*Effect of dry curing time on the moisture resistance of asphalt mixtures containing lime treated aggregates.*

however, it is felt that most of the benefit is due to the fact that the lime is held on the surface of the aggregate until the aggregate is coated with asphalt.

Additional tests involving the addition of lime slurry were conducted to evaluate curing [*8*]. The treated aggregates were cured from 0 to 90 days before mixing with asphalt. Curing was at 24°C (75°F) under both dry and wet conditions. After curing, the treated aggregates were either mixed with asphalt or were washed and then mixed with aggregate.

The level of protection increased with increased lime; however, for dry curing the effectiveness gradually decreased (Fig. 4) with increased curing time. This decrease was attributed to carbonation of the lime, which was also observed in one actual construction project. Washing of the aggregate before mixing with asphalt greatly reduced the effectiveness of the lime.

For wet curing the decrease was quite rapid (Fig. 5), possibly because of an increased rate of carbonation but also because the high humidity in the moisture room caused the lime to be removed. Washing of the moist cured, lime-coated aggregates eliminated essentially all beneficial effects of the lime (Fig. 6).

These laboratory studies as well as field experience indicate that the beneficial effects of lime are instantaneous and that curing in the stockpile is not required. In addition, the lime should be on the surface of the aggregate at the time of coating with asphalt. Also, the lime is not effective if it carbonates before mixing with the asphalt.

Comparison of Type of Lime—A study involving one asphalt and two aggregates, one of which was moisture susceptible, was conducted, using both a

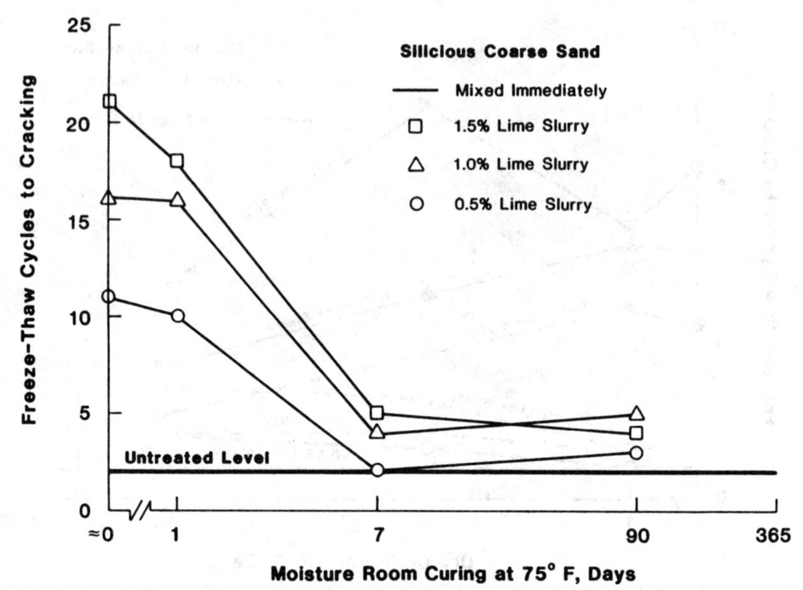

FIG. 5—*Effect of wet curing time and washing lime treated aggregates before mixing with asphalt.*

hydrated dolomitic lime (Type N) and a hydrated calcitic lime for treatment [*8*, *10*]. The mixtures were treated with 1% hydrated lime slurry and subsequently tested using the Texas boiling test. Both the hydrated calcitic and hydrated dolomitic lime (Type N) and a hydrated calcitic lime for treatment [*8*,*10*]. The mixtures were treated with 1% hydrated lime slurry and subtaining the nonmoisture-susceptible aggregate showed no adverse effects of the lime treatment.

Field Study

A field study was conducted to evaluate methods of treating asphalt mixtures with both dry hydrated lime and hydrated lime slurry in order to minimize moisture damage. Both batch and drum plants were used to produce the treated and untreated field mixtures. In addition, laboratory mixtures were prepared using methods intended to simulate field procedures.

Both dry lime and lime slurry produced improved moisture resistance of both the laboratory and field mixtures. The lime slurry, however, produced slightly better results. The only technique that was judged not to be beneficial involved injecting dry hydrated lime into the drum mixer just before the asphalt was introduced. This lack of improvement was due to the fact that the hydrated lime was lost before mixing with asphalt. It was also concluded that the wet-dry indirect tensile test, Texas boiling test, the Texas freeze-thaw pedestal test were effective in identifying moisture susceptible aggregates. The results obtained are contained in four reports [*5*,*21*–*23*] and are summarized in Ref *24*.

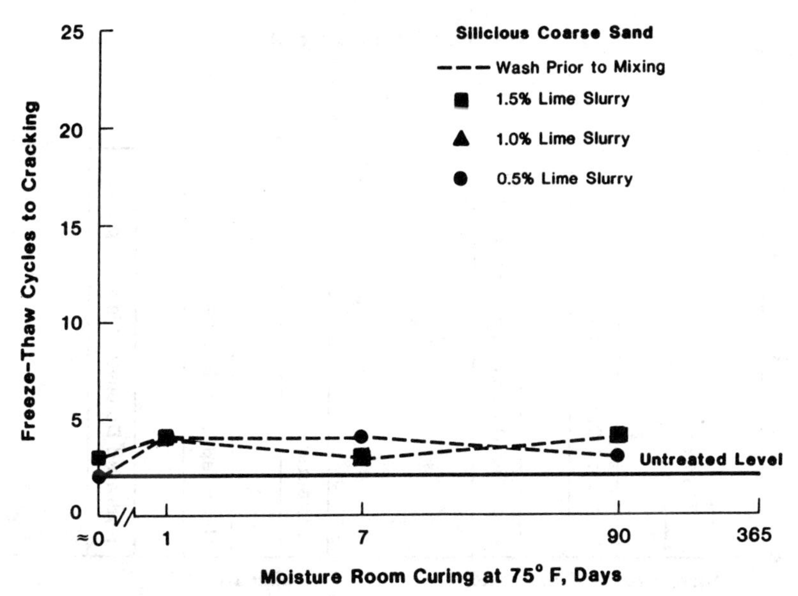

FIG. 6—*Effect of wet curing time on the moisture resistance of asphalt mixtures containing lime treated aggregates.*

Recommendations for Field Application of Hydrated Lime

Both dry lime and hydrated lime slurry can be utilized as an antistripping additive and applied to the aggregate. Lime slurry or lime in the presence of water appears to be the most effective when applied to the aggregate. It should be noted that stiffening of the mixture may occur for cooler conditions, long haul distances, and delayed placing. This may occur as surface crusting or as stiffening of the entire mixture.

The following summarizes the various techniques that can be used.

Dry Hydrated Lime

The primary problem with the addition of dry lime is holding the lime on the surface of the aggregate until it is coated with asphalt. The loss of dry lime will be greater in drum mixers, which tend to pick up the lime in the gas flow. In addition, a portion of the dry lime may be mixed into the asphalt, thus acting as a filler.

Batch and Drum Mix Plants—In both batch and drum plants, dry lime can be added to the aggregates as follows:

On the aggregate cold feed—Mixing and coating of the aggregates will be minimized. Passing the aggregate and lime through a scalping screen can improve mixing but at the same time may produce dusting and the loss of lime.

In a premixing pugmill—This technique will maximize the coating of the aggregates, but lime may be lost because of dusting.

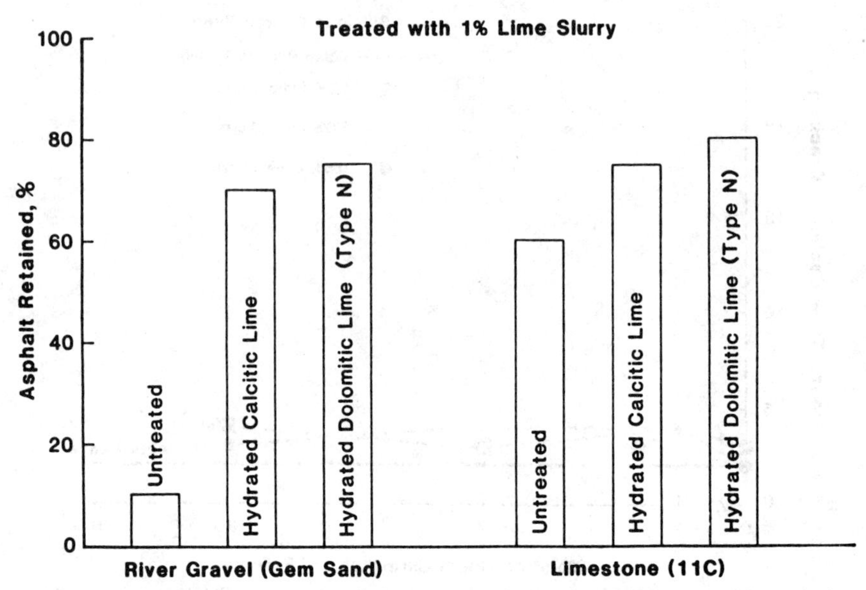

FIG. 7—*Comparison of the effect of calcitric and dolomitic hydrated lime as an antistripping additive for asphalt mixtures.*

Prior to stockpiling—This technique probably requires that the lime be added before construction of the stockpile, either by pugmilling at the plant site or by having the aggregate supplier add the lime. A large portion of the lime will probably be lost before construction because of segregation, dusting, rainfall, and so forth. This method is not recommended.

Batch Plants: In the pugmill—This technique probably maximizes mixing and coating of the aggregates and minimizes losses caused by dusting. A portion of the lime, however, may be lost in the asphalt cement.

Drum Mix Plants: In the drum before adding asphalt—This technique is not recommended unless new equipment and techniques are developed that will insure that the lime is not lost. Such modifications have been implemented by Georgia [*25*].

Lime Slurry

The primary problem with the use of lime slurry is that the water added to the aggregates must be removed by drying, thus increasing fuel costs and reducing production rates. Application techniques should minimize the amount of water, which must be removed when the aggregate enters the dryer or the drum mixer.

The lime slurry should be prepared with a minimum of approximately 30% lime with 70% water by weight in order to minimize the amount of water added to the aggregate.

Batch and Drum Mix Plants: On the cold feed—Mixing and coating of the aggregates are minimized. Passage through a scalping screen may improve aggregate coating, but the mixture can foul the screens. Since mixing is minimized, it may be possible to treat only certain aggregates by arranging the cold feed bins to place the aggregate to be treated on top of the cold feed or by treating the aggregate under the cold feed bins immediately after it is placed on the cold feed belt.

In a premixing pugmill—This method provides better coverage of the aggregate and allows a portion of the water to drain.

Prior to stockpiling—This method allows much of the water to drain, thus minimizing required drying, and would allow only one aggregate to be treated. However, it maximizes the chances of carbonation and the loss of lime. The length of time permitted in the stockpile is not well established; however, tentatively, it is recommended that stockpiling be limited to ten days or less.

Drum Mix Plants: On the slinger belt—This method minimizes the amount of mixing and coating of the aggregates and maximizes the amount of moisture that must be removed.

Dry Lime with Water

Another technique involves adding dry hydrated lime to wet aggregates or adding dry lime to dry aggregates and then spraying a small quantity of water onto the mixture. All techniques and recommendations pertaining to lime slurry also pertain to the application of dry lime and water. In general, it is felt that the water should be added to the aggregate before the dry lime is added, to minimize removal of the lime on the aggregate surface.

Hot Lime Slurry

The use of quick lime that is hydrated and slurried on the plant site offers a number of advantages. First, quick lime normally costs about the same as the cost of hydrated lime, but when slaked, it will result in about 20% more hydrated lime. In addition, slaking with excess moisture will produce a slurry with a temperature of about 82°C (180°F), which may maximize evaporation losses and the reactivity of the hydrated lime. All techniques and recommendations pertaining to lime slurry also pertain to the use of hot lime slurry.

Recommendations

The final decision as to how lime should be added should, for the most part, be left to the contractor in order to minimize costs and disruptions to the production cycle, providing that tests of the produced mixture indicate that the desired resistance to stripping was achieved. It is recommended that the lime be added as a slurry or at least with a small amount of water. However, dry lime can be used if proper precautions are taken to prevent loss of the lime before mixing with the asphalt cement. The use of a pugmill in the cold feed system is recommended since it will maximize coverage. Nevertheless, the final decision should be based on relative effectiveness and cost.

References

[*1*] McGennis, R. B., Kennedy, T. W., and Machemehl, R. B., "Stripping and Moisture Damage in Asphalt Mixtures," Research Report 253-1, Center for Transportation Research, Bureau of Engineering Research, The University of Texas at Austin, Sept. 1984.

[*2*] Kennedy, T. W., Roberts, F. L., Lee, K. W., and Anagnos, J. N., "Texas Freeze-Thaw Pedestal Test for Evaluating Moisture Susceptibility for Asphalt Mixtures," Research Report 253-3, Center for Transportation Research, Bureau of Engineering Research, The University of Texas at Austin, Aug. 1981.

[*3*] Kennedy, T. W. and Anagnos, J. N., "Lime Treatment of Asphalt Mixtures," Research Report 253-4, Center for Transportation Research, Bureau of Engineering Research, The University of Texas at Austin, July 1983.

[*4*] Kennedy, T. W., Roberts, F. L., and Anagnos, J. N., "Texas Boiling Test for Evaluating Moisture Susceptibility of Asphalt Mixtures," Research Report 253-5, Center for Transportation Research, Bureau of Engineering Research, The University of Texas at Austin, April 1983.

[*5*] Kennedy, T. W. and Anagnos, J. N., "A Field Evaluation of Techniques for Treating Asphalt Mixtures with Lime," Research Report 253-6, Center for Transportation Research, Bureau of Engineering Research, The University of Texas at Austin, Nov. 1984.

[*6*] Kennedy, T. W. and Anagnos, J. N., "Modified Test Procedure for Texas Freeze-Thaw Pedestal Test," Research Report 253-7, Center for Transportation Research, Bureau of Engineering Research, The University of Texas at Austin, Nov. 1984.

[*7*] Kennedy, T. W. and Anagnos, J. N., "Indirect Tensile Test for Evaluating Moisture Susceptibility of Asphalt Mixtures," Research Report 253-8, Center for Transportation Research, Bureau of Engineering Research, The University of Texas at Austin, Nov. 1984.

[*8*] Kennedy, T. W. and Anagnos, J. N., "Techniques for Reducing Moisture Damage in Asphalt Mixtures," Research Report 253-9F, Center for Transportation Research, Bureau of Engineering Research, The University of Texas at Austin, Nov. 1984.

[*9*] Russam, K. and Coleman, J. D., "The Effect of Climatic Factors on Subgrade Moisture Conditions," *Geotechnique,* Vol. 11, 1961.

[*10*] Kennedy, T. W., "Use of Hydrated Lime in Asphalt Mixtures," National Lime Association, Washington, DC, March 1984.

[*11*] The Asphalt Institute, "Cause and Prevention of Stripping in Asphalt Pavements," Educational Series No. 10 (ES-10), Jan. 1981.

[*12*] Kennedy, T. W., Roberts, F. L., and Lee, K. W., "Evaluation of Moisture Susceptibility of Asphalt Mixtures Using the Texas Freeze-Thaw Pedestal Test," *Proceedings of the Association of Asphalt Paving Technologists,* Vol. 51, Feb. 1982, pp. 327–341.

[*13*] Kennedy, T. W., Roberts, F. L., and Lee, K. W., "Evaluating Moisture Susceptibility of Asphalt Mixtures Using the Texas Boiling Test," Transportation Research Record, Transportation Research Board, National Academy of Sciences, Washington, DC, 1983.

[*14*] Kennedy, T. W. and Anagnos, J. N., "A Method for Conducting the Static and Repeated-Load Indirect Tensile Tests," Research Report 183-14, Center for Transportation Research, Bureau of Engineering Research, The University of Texas at Austin, Aug. 1983.

[*15*] Tunicliff, D. G. and Root, R. E., "Testing of Asphalt Concrete for Effectiveness of Antistripping Additives," *Proceedings of the Association of Asphalt Paving Technologists,* Vol. 52, Feb. 1983, pp. 535–553.

[*16*] "Prediction of Moisture Induced Damage to Bituminous Paving Mixtures," Manual of Testing Procedures, Texas State Department of Highways and Public Transportation, Austin, TX, Aug. 1984.

[*17*] Plancher, H., Miyake, G., Venable, R. L., and Petersen, J. C., "A Simple Laboratory Test to Indicate the Susceptibility of Asphalt-Aggregate Mixtures to Moisture Damage During Repeated Freeze-Thaw Cycling," *Proceedings of the Canadian Technical Asphalt Association,* Vol. 25, Nov. 1980.

[*18*] Shah, S. C., "Antistripping Additives in Lieu of Mineral Filler in Asphaltic Concrete Mixtures," Research Report 88, Research Project 72-3b(B), Louisiana HPR 1(12), Louisiana Department of Transportation, Baton Rouge, LA, April 1975.
[*19*] "Virginia Test Method for Heat Stable Additives," VTM-13, Virginia Department of Transportation, Richmond, VA, 1978.
[*20*] "Qualifications of Antistripping Additives," DODT Designation TR 317-77, Louisiana Department of Transportation, Baton Rouge, LA, Feb. 1977.
[*21*] Turnham, N., "The Effects of Lime as an Antistripping Agent," Research Report Number SS24.1, District 17, Texas State Department of Highways and Public Transportation, Austin, TX, Feb. 1983.
[*22*] Button, J. W. and Epps, J. A., "Evaluation of Methods of Mixing Lime in Asphalt Paving Mixtures," Report RF 4773-1, prepared for the Texas Hot Mix Association, Texas Transportation Institute, Texas A&M University, College Station, TX, July 1983.
[*23*] Button, J. W., Kennedy, T. W., Epps, J. A., and Turnham, N., "A Field Study Using Lime as an Antistripping Additive for Asphalt Pavements," Research Report THM-1F, Texas State Department of Highways and Public Transportation, Center for Transportation Research at The University of Texas at Austin, and Texas Transportation Institute at Texas A&M University, July 1983.
[*24*] Kennedy, T. W., Turnham, N., Jr., Epps, J. A., Smoot, C. W., Young, F. M., Button, J. W., and Zeigler, C. D., "Evaluation of Methods for Field Applications of Lime to Asphalt Concrete Mixtures," *Proceedings of the Association of Asphalt Paving Technologists,* Vol. 52, Feb. 1983, pp. 508–528.
[*25*] Collins, R., "Georgia's Experience with the Use of Hydrated Lime in Asphalt Concrete Mixtures," Southern Association of State Highway and Transportation Officials Meeting, Mobile, AL, 1982.

Joe W. Button [1]

Maximizing the Beneficial Effects of Lime in Asphalt Paving Mixtures

REFERENCE: Button, J. W., "**Maximizing the Beneficial Effects of Lime in Asphalt Paving Mixtures,**" *Evaluation and Prevention of Water Damage of Asphalt Pavement Materials, ASTM STP 899,* B. E. Ruth, Eds., American Society for Testing and Materials, Philadelphia, 1985, pp. 134–146.

ABSTRACT: Laboratory and field tests were conducted to evaluate the use of hydrated lime as an antistrip additive in hot mix asphalt concrete. Batch and drum mix plants were used to prepare the paving mixtures. Lime was added dry and in slurry form. Individual aggregates and the total aggregate were separately treated with lime slurry and allowed to age for different time periods from a few minutes to 30 days before mixing with asphalt. Laboratory mixed and plant mixed asphalt concrete was tested using indirect tension and resilient modulus before and after moisture conditioning. Results indicate that lime is effective in reducing moisture susceptibility and that it is most effective when applied in the presence of moisture. In addition, a time delay after application of lime to aggregate is unnecessary. There are no significant differences in mixtures produced in batch and drum plants.

KEY WORDS: pavements, lime, additives, tensile strength, asphalt mixtures, water susceptibility, antistripping, resilient modulus, hydrated lime

Introduction

Hydrated Lime in Asphalt Concrete

In the 1910 to 1930 era there were applications of 1 to 2% hydrated lime in asphalt paving mixtures [*1*]. A few states even specified hydrated lime in their specifications. It provided a combination antistripping agent and mineral filler. There are no statistics on the extent of this relatively small use; and being insignificant, it was never promoted by the lime industry [2]. Unaccountably, this use almost disappeared from 1930 to about 1957. Presumably, since the use of mineral dust fillers was generally advocated and specified by state highway departments, asphalt contractors chose the less expensive fillers. Also, numerous

[1] Associate research engineer, Texas Transportation Institute, Texas A&M University, College Station, TX 77843.

specialty organic antistripping compounds were vigorously promoted and largely replaced hydrated lime in spite of their higher cost [1].

Field performance information together with recently obtained laboratory test results that used advanced testing techniques have confirmed that lime is one of the most effective chemical antistrip agents available to the pavement design engineer [3–5]. In addition, lime has been shown to reduce the oxidative hardening of asphalt cement [6–10]. The application of a lime slurry to the aggregate some time before producing the asphalt concrete appears to be the best method for insuring that the lime acts as an effective antistrip chemical [3,4]. Some specifications require a delay of 30 days between application of the slurry to the aggregate and hot mixing with asphalt. Other specifications do not require a specific time delay, and still other specifications allow lime to be applied in the dry form. Methods of applying lime and the point in the paving mixture production stream at which lime is applied also vary widely.

Objectives of Study

Standardized methods for applying lime need to be defined based on an understanding of plant operations and equipment capabilities as well as the influence that the method of application may have on the effectiveness of the lime as an antistrip agent. Costs associated with applying lime by the different techniques need to be considered.

This report describes the findings of a field study to evaluate techniques for adding dry lime and slurried lime in batch and drum mix plants [9,10]. The primary objectives of the research program are to (1) determine the effectiveness of lime as an antistrip additive when added either dry or in slurry, (2) investigate the effect of time delay after lime treatment of aggregate, (3) evaluate the point of entry of lime in the production system, and (4) assess the differences in mixtures produced in the batch and drum mix plants.

Materials

Asphalt Cement

AC-20 asphalt cement from the Exxon refinery in Baytown, TX, was used throughout this study. It was produced by the propane deasphalting process. Properties of the original asphalt are given in Table 1.

Aggregates

Pea gravel, washed sand, and field sand were combined to produce the project gradation. The pea gravel and washed sand consist of subrounded smooth textured particles. All three of the aggregates are siliceous. Gradations of the individual aggregates, the project gradation, percentages of each aggregate combined, specific gravities and the specification are given in Table 2.

TABLE 1—*Properties of original asphalt cement.*

Characteristic Measured	Measurement
Viscosity	
77°F (25°C), poise	2.75 × 10^6
140°F (60°C), poise	1983
275°F (135°C), poise	3.78
Penetration	
77°F (100 gm, 5 s)	60
39.2°F (4°C) (100 g, 5 s)	0
39.2°F (4°C) (200 g, 60 s)	12
Softening point, °C	50 (122°F)
Flash point, °C	315+ (600+°F)
Specific gravity	1.03
After thin film oven test	
viscosity at 60°C (140°F)	5316
penetration at 25°C (77°F)	31
weight loss, %	0[a]
ductility, cm	150+
viscosity ratio	2.68
retained penetration, %	52

[a]Actually a slight gain in weight (0.07%) was indicated by repeated tests.

TABLE 2—*Aggregate gradations and specific gravities.*[a]

Gradations and Specific Gravity	Processed Pea Gravel		Washed Sand		Field Sand		Combined Gradation	Specification
Gradation, %								
plus ½ in.	0		...		...		0	0
½ to ⅜ in.	0.4		0		...		0.3	0 to 5
⅜ to No. 4	61.6		3.6		...		38.7	20 to 50
No. 4 to No. 10	35.5		18.7		...		24.8	10 to 30
plus No. 10	97.5		22.3		0		63.8	50 to 70
No. 10 to No. 40	2.0		49.3		0.4		8.7	0 to 30
No. 40 to No. 80	0.2		26.2		48.0		15.0	4 to 25
No. 80 to No. 200	0.1		1.7		44.8		10.7	3 to 25
minus No. 200	0.2		0.5		6.8		1.8	0 to 6
Percent combined	62	+	15	+	23	=	100	
Bulk specific gravity								
plus No. 10	2.639		2.615		...		...	
minus No. 10 to plus No. 80	...		2.637		2.637		...	
minus No. 80	...		...		2.709		...	
combined	...		...		...		2.646	

[a]Data furnished by Texas SDHPT District 17 laboratory personnel.

Lime

Dry hydrated lime was supplied in bags. When used as a slurry, it was mixed in a slurry mixer at a 70 to 30 weight ratio of water to lime. Lime (dry or in slurry) was added at a rate of 1.5% of dry lime by weight of the aggregate treated.

Asphalt Paving Mixture

The hot mixed asphalt concrete (HMAC) mixture used in this study met the Texas State Department of Highways and Public Transportation (SDHPT) specifications for a fine graded surface course. Texas SDHPT personnel designed the mixture without hydrated lime using their standard design procedures. Laboratory test results from this mixture design are given below [*11*]:

- Asphalt content, 5%.
- Average density, 95.5% of theoretical maximum.
- Air-void content, 4.5%.
- Hveem stability, 41.
- Cohesiometer value, 164.

Experimental Program

Scope of Investigation

The field study was performed on a pavement reconstruction and widening project. Hydrated lime was added to the HMAC using 13 different methods. Lime was added with no adjustment to mixture design. Individual pavement materials (asphalt and aggregates), paving mixtures from the batch and drum mix plants, and pavement cores were obtained and tested in the laboratory.

Laboratory Mixed/Laboratory Compacted Mixtures

Samples of the project asphalt cement, aggregates, and hydrated lime were obtained. These materials were used to prepare and compact paving mixtures in the laboratory.

Laboratory mixed and compacted specimens were fabricated by applying the lime by seven different methods. One and one-half percent lime by weight of aggregate was added dry or in slurry to fine and coarse aggregates individually and to the total aggregate. Selected lime-treated aggregates were allowed to cure for 2 or 30 days before mixing with asphalt. One set of specimens was made by adding 1.5% silica flour to the total aggregate in an attempt to determine the effects of merely adding an "inert" filler instead of lime.

Field Mixed/Laboratory Compacted Mixtures

Samples of the field mixtures were obtained from the asphalt mixing plant. They were immediately transported to the laboratory and compacted using a gyratory molding press to fabricate 102-mm (4-in.) diameter briquettes. Reheating was necessary to maintain a compaction temperature of 121°C (250°F), but reheating was kept to a minimum in order to minimize any changes in mixture properties.

Methods of adding hydrated lime to the field mixtures and identification of the codes used in several subsequent figures are given in Table 3.

TABLE 3—*Explanation of codes used on figures for field mixtures.*

Method of Adding Lime	Code Used on Figures	
BATCH PLANT		
Control (no lime)	C_b	
Dry lime in pugmill[a]	dry	
Slurry on total aggregate + 2*da* in stockpile	*S*	
DRUM MIX PLANT		
Control (no lime)	C_d	
Dry lime on total aggregate at cold feed belt	CFB	dry
Dry lime at center of drum thru fines feeder	CD	dry
Slurry on field sand at cold feed belt	FS	slurry
Slurry on washed sand at cold feed belt	WS	slurry
Slurry on pea gravel at cold feed belt	PG	slurry
Slurry on total aggregate at cold feed belt	TA	slurry
Slurry on field sand + 2*da* in stockpile	FS	2-day slurry
Slurry on washed sand + 2*da* in stockpile	WS	2-day slurry
Slurry on pea gravel + 2*da* in stockpile	PG	2-day slurry
Slurry on total aggregate + 2*da* in stockpile	TA	2-day slurry
Slurry on total aggregate + 30*da* in stockpile	TA + 30 day	

[a]Dry lime was added and mixed for 20 s before addition of asphalt cement.

Test Results and Discussion

Laboratory Mixed/Laboratory Compacted Mixtures

In this phase of the test program, only the fine and coarse aggregates were treated with lime slurry. Fine aggregate consisted of the combined field sand and washed sand. Coarse aggregate consisted of the pea gravel. One and one-half percent lime by weight of total aggregate was added to these mixtures whenever the coarse and fine or total aggregate was treated. Test results from this phase of the work are given on Table 4.

Resilient Modulus—Resilient moduli at 25°C (77°F) were determined before and after vacuum saturation and soaking in water for seven days at 25°C.

Ratios were computed by dividing resilient modulus after moisture treatment by its corresponding original value before moisture treatment. For example

$$\text{resilient modulus ratio} = M_R \text{ after moisture treatment} / M_R \text{ of original specimen}$$

Most of the resilient modulus ratios are greater than 1.0, which indicates higher values of resilient modulus after moisture treatment than before. This unlikely phenomenon may be due to evaporative cooling of the saturated specimens during testing.

Observation of the resilient modulus ratios in Fig. 1 reveals that those mixtures treated with lime slurry consistently yield the highest values. This is consistent with previous laboratory test results [*4*]. Those mixtures treated with slurry two days or more before mixing and compacting yielded higher resilient modulus ratios than those treated only five minutes before mixing and compacting.

TABLE 4—*Mean values of data from laboratory mixed and compacted specimens[a] before and after freeze-thaw.*

Test Procedure	Mixture Type								
	Control	Silica Flour	Dry Lime	Lime in Asphalt Cement	Slurry on Total Aggregate 5 min	Slurry on Fine Aggregate 2-days	Slurry on Coarse Aggregate 2-days	Slurry on Total Aggregate 2-days	Slurry on Total Aggregate 30-days
	C	SF	DL	LA	ST5	SF2	SC2	ST2	ST30
Overall average									
Air-void content, %[b]	7.3	7.2	7.5	9.1	5.6	7.4	7.4	6.0	7.9
Resilient modulus, psi × 10^3									
−13°F	1980 (7.3)	1860	1 990 (6.8)	1790 (8.9)	1 930 (4.8)	2 430 (7.3)	2070 (7.4)	2 080 (5.7)	1930
33°F	1650	1780	1 860	1480	1 590	1 670	1490	1 370	1480
68°F	493	467	550	363	505	370	381	511	343
104°F	44	36	50	29	37	25	27	40	32
77°F[b]	271 (7.3)	234	344 (7.2)	230 (9.1)	283 (5.6)	217 (7.4)	201 (7.4)	261 (6.0)	182
77°F (after freeze-thaw)	245 (7.5)	208	315 (8.4)	235 (9.7)	329 (7.2)	315 (7.1)	286 (7.3)	304 (6.5)	283
77°F (after soak)	281 (7.4)	280	391 (7.4)	255 (8.8)	362 (5.2)	291 (7.7)	277 (7.7)	371 (5.7)	298
Marshall test									
stability, lb	280 (7.1)	260	290 (7.1)	140 (8.8)	310 (5.1)	180 (7.2)	200 (7.1)	330 (5.7)	230
flow, 0.01 in.	12	14	19	15	17	17	18	16	15
stab (after soak)	...	450	430 (7.4)	240 (8.8)	490 (5.2)	320 (7.7)	360 (7.7)	610 (5.7)	320
flow (after soak)	...	15	19	18	18	19	18	18	15
Hveem stability									
stability	19 (7.4)	19	16 (7.4)	15 (8.8)	14 (5.2)	14 (7.7)	14 (7.7)	16 (5.7)	15
stab (after soak)	25	23	18	20	17	19	19	20	19
Splitting tensile test[c]									
stress, psi	110	110	110 (6.8)	80 (8.9)	110 (4.8)	90 (7.3)	90 (7.4)	120 (5.7)	100
strain, in./in.	0.012 (7.3)	0.012	0.010	0.011	0.010	0.011	0.012	0.010	0.014
secant Mod, psi	9400	8900	10 600	6800	11 500	7 600	7500	12 200	6600
stress (after freeze-thaw)	90	90	90 (8.4)	70 (9.7)	120 (7.2)	110 (7.1)	110 (7.3)	120 (6.5)	100
strain (after freeze-thaw)	0.013 (7.5)	0.013	0.010	0.009	0.010	0.010	0.012	0.009	0.013
secant Mod (after freeze-thaw)	7400	7100	9 700	8000	11 500	10 800	9300	14 200	8200

[a]Each value represents an average for three tests. Numbers within the table in parentheses indicate average air-void content of the three specimens tested to produce the data.

[b]Each value represents an average for twelve tests.

[c]Values were obtained at point of specimen failure.

NOTE: 1 psi = 6894 Pa, 1 lb = 0.4536 kg, and 1 in. = 0.0254 m.

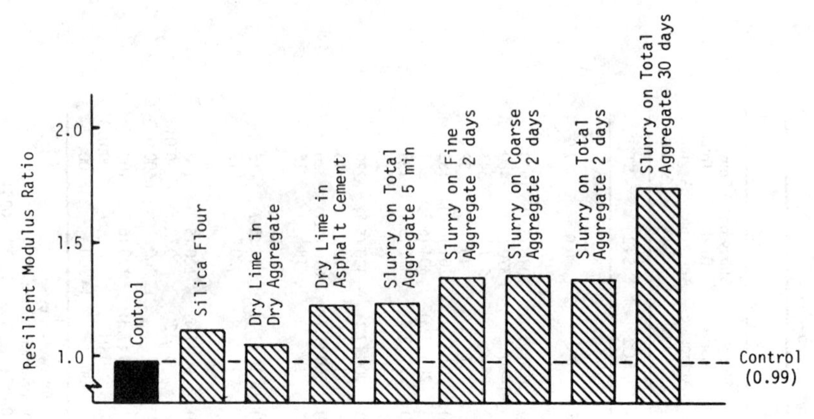

FIG. 1—*Resilient modulus ratios for seven-day soak moisture treatment on laboratory mixed/laboratory compacted specimens (measured at 25°C (77°F).*

Tensile Properties—The indirect tension test was conducted at a temperature of 25°C (77°F) and a deformation rate of 51 mm (2-in.)/min. Cylindrical specimens 102 mm (4-in.) in diameter were tested before and after moisture treatment using the accelerated Lottman [*12*] freeze-thaw procedure.

Tensile strength ratios are plotted on Fig. 2. Notably higher tensile strength ratios are exhibited by those mixtures treated with slurry. This indicates that, of those methods tested, slurry is the most beneficial form of lime application.

The most frequently reported theory regarding the ability of lime to decrease moisture-induced damage in asphalt concrete involves direct contact of wetted lime on the aggregate surface. Lime is purported to alter the surface chemistry of the aggregate thus producing a more tenacious bond between the asphalt and the aggregate. However, in this experiment, Mixtures DL and LA were made without any water, and both of them exhibited improved resistance to moisture (Fig. 2). Therefore, other mechanisms appear to be involved.

From the standpoint of tensile strength retention, there appears to be no significant advantage to aging the lime treated aggregate before mixing with asphalt.

Field Mixed/Laboratory Compacted Mixtures

Summaries of test data from this portion of the work are given in Tables 5 and 6.

Tensile Properties—Figure 3 indicates that tensile strength ratios are considerably higher for those mixtures containing lime. Furthermore, those mixtures containing lime slurry generally yielded higher retained strength than those containing dry lime. Mixture CFB (dry lime on cold feed belt), however, yielded a tensile strength ratio of the same order of magnitude as those mixtures containing lime slurry. It should be noted that the aggregate on the cold feed belt was moist. Mixture CD (dry lime through fines feeder) gave comparatively low ratios for

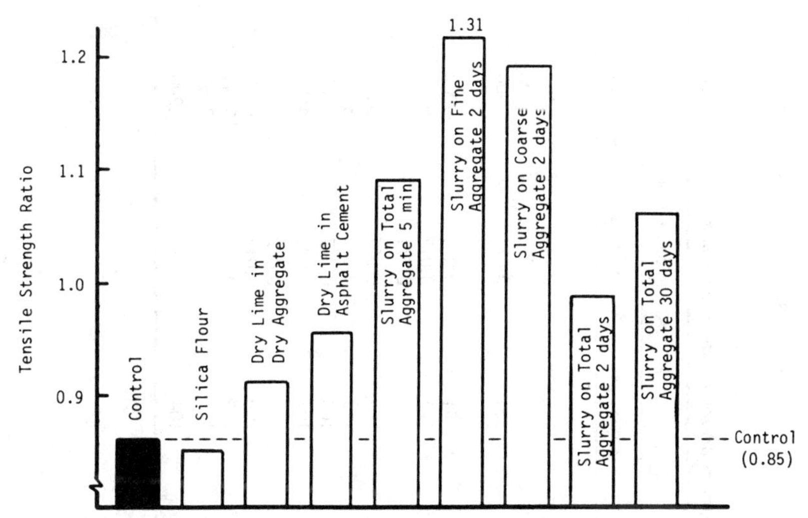

FIG. 2—*Tensile strength ratio for laboratory mixed and compacted specimens after freeze-thaw treatment.*

both tensile strength (Fig. 3) and resilient modulus. These results are in agreement with results on tests of identical mixtures conducted at The University of Texas [*13*]. This indicates a possible disadvantage when dry lime is added through the fines feeder into the center of the drum. It is possible that a portion of the lime was blown out of the drum during this experiment and deposited in the baghouse. Furthermore, considerable drying of the aggregate has occurred by the time it reaches this portion of the drum and reaction time of the lime and aggregate before contact with asphalt is almost nil. It should be pointed out, however, that subjective stripping test results on these mixtures reported by Turnham [*11*] showed little difference between Mixtures CFB and CD.

There were no detectable differences in tensile strength ratio when the aggregate was treated with lime slurry immediately before mixing or stockpiled for two days. Tensile strength ratios of resulting mixtures, did, however, appear to decrease when slurry-treated aggregate was stockpiled for 30 days.

There are no measurable differences in tensile properties or water susceptibility between mixtures made in the batch plant and those made in the drum mix plant.

Conclusions and Recommendations

Conclusions

1. Hydrated lime is effective in reducing moisture-induced damage of the paving mixture considered in this study.

2. The most effective method for applying lime is in the presence of moisture. The moisture may be on the surface of the aggregate to be treated or the lime may be introduced as a slurry.

TABLE 5—*Summary of data from splitting tension test at 77°F of field mixed/lab compacted mixtures before and after Lottman moisture treatment.*

	Original			After Freeze-Thaw		
Method of Adding Lime	Stress, psi	Strain, in./in.	Modulus, psi	Stress, psi	Strain, in./in.	Modulus, psi
		BATCH PLANT				
Control (no lime)	83	0.018	4500	83	0.017	5100
Dry lime in pugmill	78	0.020	3900	89	0.016	5800
Slurry on total aggregate + 2*da* in stockpile	100	0.017	5700	134	0.016	8300
		DRUM MIX PLANT				
Control (no lime)	104	0.016	6500	81	0.009	9000
Dry lime on cold feed belt	82	0.018	4500	103	0.017	6300
Dry lime at center of drum through fines feeder	87	0.016	5300	63	0.011	5700
Slurry on field sand at cold feed belt	86	0.014	6300	105	0.017	6400
Slurry on washed sand at cold feed belt	96	0.015	6500	122	0.015	8200
Slurry on pea gravel at cold feed belt	81	0.018	4500	103	0.017	6200
Slurry on total aggregate at cold feed belt	80	0.017	4700	101	0.021	4800
Slurry on field sand + 2*da* in stockpile	85	0.018	4800	101	0.012	6000
Slurry on washed sand + 2*da* in stockpile	87	0.018	5000	109	0.017	6500
Slurry on pea gravel + 2*da* in stockpile	88	0.017	5300	119	0.017	7100
Slurry on total aggregate + 2*da* in stockpile	91	0.016	5800	123	0.018	7300
Slurry on total aggregate + 30*da* in stockpile	98	0.014	6900	104	0.013	8200

NOTE: 1 psi = 6894 Pa, in./in. = m/m.

TABLE 6—*Average values of resilient modulus of field mixed/lab compacted mixtures before and after treatment with moisture.*

Method of Adding Lime	Resilient Modulus, psi × 10^3							Overall Average Air Voids, %
	−23°C (−10°F)	1°C (33°F)	20°C (68°F)	40°C (104°F)	25°C (77°F)	77°F, after Freeze-Thaw	77°F, after 7-Day Soak	
Batch Plant								
Control (no lime)	2080	1570	290	24	140	110	210	8.3
Dry lime in pugmill	2120	1550	300	22	150	160	220	9.8
Slurry on total aggregate + 2*da* in stockpile	2070	1750	410	27	200	265	377	5.6
Drum Mix Plant								
Control (no lime)	1970	1890	470	27	240	240	390	7.8
Dry lime on cold feed belt	2080	1490	310	22	160	190	340	7.2
Dry lime at center of drum through fines feeder	1760	1740	390	20	200	171	270	8.3
Slurry on field sand at cold feed belt	1810	1590	330	20	180	200	290	6.9
Slurry on washed sand at cold feed belt	1860	1640	400	27	220	250	370	6.9
Slurry on pea gravel at cold feed belt	1920	1690	330	21	170	200	310	8.0
Slurry on total aggregate at cold feed belt	1980	1720	340	22	190	170	400	6.3
Slurry on field sand + 2*da* in stockpile	1740	1670	350	20	170	180	280	6.0
Slurry on washed sand + 2*da* in stockpile	2180	1690	340	19	170	200	250	7.1
Slurry on pea gravel + 2*da* in stockpile	1940	1720	460	21	190	250	350	7.6
Slurry on total aggregate + 2*da* in stockpile	1990	1790	460	22	210	230	400	7.2
Slurry on total aggregate + 30*da* in stockpile	2110	1260	200	16	120	150	190	9.1

NOTE: 1 psi = 6894 Pa.

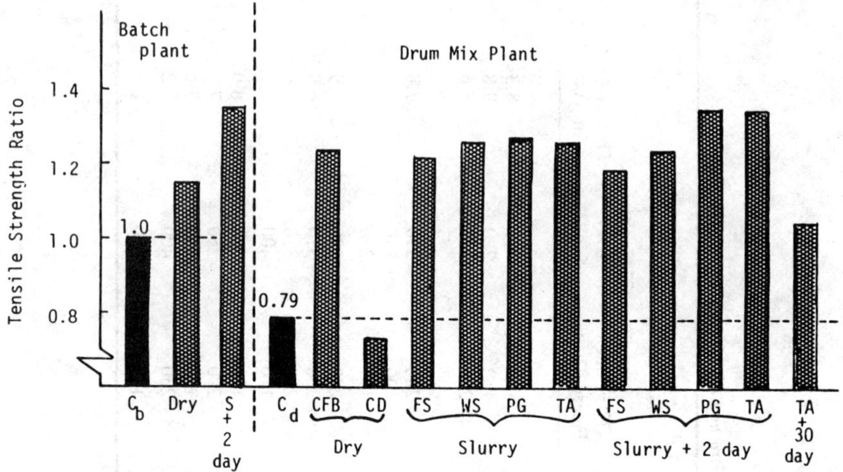

FIG. 3—*Tensile strength ratios before and after freeze-thaw moisture treatment for field mixed/laboratory compacted mixtures.*

3. Dry lime added through a fines feeder system in a drum mix plant was found to be ineffective as an antistrip additive.

4. Dry lime added in the pugmill of a batch plant was effective in improving resistance to moisture damage; however, lime slurry applied to the aggregate before entry into the batch plant was more effective.

5. Lime is an effective antistrip additive in both batch and drum mix plants. In fact, there were no significant differences in properties of the mixtures prepared in the batch plant and those prepared in the drum plant.

6. It appears that significantly less lime could have been used to produce the antistripping qualities of this paving mixture particularly if moisture and mixing had been optimized.

7. Application of lime as an antistrip additive may allow the use of otherwise unacceptable materials and thereby reduce materials and transportation costs.

8. Ratios of tensile strength and resilient modulus before and after treatment with moisture appear to be satisfactory laboratory procedures for estimating resistance to moisture damage of asphalt concrete mixtures, however, moisture conditioning of test specimens must be carefully controlled such that all specimens are treated uniformly.

9. The mixture containing 1.5% silica flour instead of lime generally exhibited properties similar to the control mixture.

Recommendations

1. When lime is to be used in HMAC, it should be included in the design phase.

2. In order to maximize the beneficial effects of lime, thorough mixing of the lime with the aggregate should be accomplished before entry of the treated aggregate into the plant.

3. If a given aggregate(s) is known to be water susceptible and if it is desired to treat only that aggregate(s) with lime, it should be treated before placement in the cold bin or some other mixing, or stirring apparatus should be employed before combining the lime-treated aggregate with the untreated aggregate.

4. Stockpiling of lime-treated aggregate for the sole purpose of decreasing moisture susceptibility is neither necessary nor recommended. In fact, prolonged exposure of lime-treated aggregate to the elements may decrease the beneficial effects of the lime.

5. Air-void content in compacted samples and methods utilized for moisture conditioning samples must be carefully controlled when comparing strength, stiffness, and stability of asphalt paving mixtures.

Acknowledgments

This work was sponsored by the Texas Hot Mix Asphalt Pavement Association in cooperation with the Texas State Department of Highways and Public Transportation and Young Brothers, Contractors of Waco, TX.

References

[1] Boynton, R. S., *Chemistry and Technology of Lime and Limestone,* 2nd ed., Wiley, New York, 1980.

[2] National Lime Association, "Hydrated Lime in Asphalt Paving," Bulletin No. 325, Arlington, VA, 1961.

[3] Kennedy, T. W., Roberts, F. L., and Lee, K. W., "An Evaluation of Antistripping Agents for Asphalt Mixtures," Research Report 253-4, Center for Transportation Research, University of Texas at Austin, 1982.

[4] Button, J. W., Valdez, R. R., Epps, J. A., and Little, D. N., "Development Work on a Test Procedure to Identify Water Susceptible Asphalt Mixtures," Research Report 287-1, Texas Transportation Institute, Texas A&M University, College Station, TX, June 1982.

[5] Kennedy, T. W., Roberts, F. L., and Lee, K. W., "Evaluation of Moisture Susceptibility of Asphalt Mixtures Using the Texas Freeze-Thaw Pedestal Test," *Proceedings of the Association of Asphalt Paving Technologists,* Vol. 51, 1982, pp. 327–341.

[6] Petersen, J. C., "Relationships Between Asphalt Chemical Composition and Performance-Related Properties," prepared for presentation at the annual meeting of the Asphalt Emulsion Manufacturers Association, Las Vegas, NV, March 1982.

[7] Plancher, H., Green, E. L., and Petersen, J. C., "Reduction of Oxidative Hardening of Asphalts by Treatment with Hydrated Lime—A Mechanistic Study," *Proceedings of the Association of Asphalt Paving Technologists,* Vol. 45, 1976, pp. 1–24.

[8] Chachas, C. V., Liddle, W. J., Peterson, D. E., and Wiley, M. L., "Use of Hydrated Lime in Bituminous Mixtures to Decrease Hardening of the Asphalt Cement," National Technical Information Service Report No. PB-213 170, final report, Utah State Department of Highways, Materials and Test Division, Salt Lake City, UT, Dec. 1971.

[9] Button, J. W. and Epps, J. A., "Evaluation of Methods of Mixing Lime in Asphalt Paving Mixtures," Report 4773-1, Texas Transportation Institute, Texas A&M University, College Station, TX, July 1983.

[*10*] Button, J. W., "Evaluation of Methods of Mixing Lime in Bituminous Paving Mixtures in Batch and Drum Mix Plants," Master's thesis, Texas A&M University, College Station, TX, May 1984.

[*11*] Turnham, N., "The Effects of Lime as an Antistripping Agent," Report Number SS24-1, Texas State Department of Highways and Public Transportation, Austin, TX, Feb. 1983.

[*12*] Lottman, R. P., "Predicting Moisture-Induced Damage to Asphaltic Concrete: Field Evaluation Phase," National Cooperative Highway Research Program, Project 4-8(3)/1, final report, Jan. 1982.

[*13*] Kennedy, T. W., Turnham, N., Epps, J. A., Smooth, C. W., Young, F. M., Button, J. W., and Zeigler, C. D., "Evaluation of Methods for Field Applications of Lime to Asphalt Concrete Mixtures," Association of Asphalt Paving Technologists, Vol. 52, Feb. 1983, pp. 508–534.

Author Index

Subject Index

T

V–W